Praise for *Enemy of All Mankind*

"A kaleidoscopic rumination on the ways in which a single event, and the actions of a handful of men with no obvious access to the levers of state power, can change the course of history. . . . Steven Johnson treats us to fascinating digressions on the origins of terrorism; celebrity and the tabloid media; the tricky physics of cannon manufacture; and the miserable living conditions of the average seventeenth-century seaman."

—*The New York Times Book Review*

"Steven Johnson argues with verve and conviction in his thoroughly engrossing *Enemy of All Mankind*. . . . Because *Enemy of All Mankind* offers, among its many pleasures, a solid mystery story, it would be wrong to reveal the outcome. But it's surprising. So, too, are the many larger themes that Mr. Johnson persuasively draws from his seaborne marauders. It is possible that readers may not be eager to learn about seventeenth-century metallurgy, or the coalescing of nation-states, or the rise of world-wide commerce in the era; but they will be glad they did once they encounter these matters here. All the author's more surprising suppositions are not merely stapled onto the narrative but seem to have grown there effortlessly during the course of a spirited, suspenseful, economically told tale whose significance is manifest and whose pace never flags."

—*The Wall Street Journal*

"A page-turner of a book . . . we can thank Johnson for combing the archives, describing in vivid detail the life of pirates that we thought we knew—most likely through motion pictures—when in truth we didn't. . . . *Enemy of All Mankind* covers lots of territory, including the beginnings of the British Empire, and it's a good read, made all the better by Johnson's clever storytelling and an unforgettable pirate named Henry Every."

—*The Washington Post*

"It is the perfect book to cozy up to during a pandemic. . . . In addition to providing captivating 'yo ho ho and a bottle of rum' action, the author examines the geopolitical and cultural implications of Every's spasm of violence. His subject changed the very nature and geography of piracy in the eighteenth century."

—*USA Today*

"Enough adventures to fill a Netflix series . . . [Johnson] skillfully makes sweeping historical points from bloody swashbuckling details."

—*Star Tribune*

"Entertaining and erudite . . . Johnson's lucid prose and sophisticated analysis brings these events to vibrant life. This thoroughly enjoyable history reveals how a single act can reverberate across centuries."

—*Publishers Weekly* (starred review)

"This gripping tale of the misadventures of seventeenth-century pirate Henry Every sugarcoats revelations about the birth of the multinational corporation, the rise of tabloid media, and the surprising origins of democracy. Johnson's writing is lyrical and lucid, and his multidisciplinary curiosity is infectious."

—Next Big Idea Club

"Johnson is one of those polymath writers who links events and subjects most of us wouldn't see as related, always to enlightening effect . . . intriguing . . . relevant to our own world. Johnson doesn't just write about the heyday of piracy; he connects it to the growth of nation-states, the history of the first multinational corporation, the origins of democracy and the birth of the tabloid media, among other things. . . . An amazing story, but the real one Johnson tells in *Enemy of All Mankind* is even more so."

—*Tampa Bay Times*

"Johnson weaves a tapestry of treasure, tribunals, emperors, atrocities, and a pirate's life at sea . . . Consummate popular history: fast-paced, intelligent, and entertaining."

—*Library Journal*

ALSO BY STEVEN JOHNSON

Interface Culture:
How New Technology Transforms the Way We Create and Communicate

Emergence:
The Connected Lives of Ants, Brains, Cities, and Software

Mind Wide Open:
Your Brain and the Neuroscience of Everyday Life

Everything Bad Is Good for You:
How Today's Popular Culture Is Actually Making Us Smarter

The Ghost Map:
The Story of London's Most Terrifying Epidemic—and How It Changed Science, Cities, and the Modern World

The Invention of Air:
A Story of Science, Faith, Revolution, and the Birth of America

Where Good Ideas Come From:
The Natural History of Innovation

Future Perfect:
The Case for Progress in a Networked Age

How We Got to Now:
Six Innovations That Made the Modern World

Wonderland:
How Play Made the Modern World

Farsighted:
How We Make the Decisions That Matter the Most

Extra Life:
.A Short History of Living Longer

ENEMY of ALL MANKIND

A True Story of
Piracy, Power, and History's
First Global Manhunt

STEVEN JOHNSON

RIVERHEAD BOOKS
NEW YORK

RIVERHEAD BOOKS
An imprint of Penguin Random House LLC
penguinrandomhouse.com

The Library of Congress has catalogued the Riverhead hardcover edition as follows:
Names: Johnson, Steven, 1968– author.
Title: Enemy of all mankind: a true story of piracy, power, and history's
first global manhunt / Steven Johnson.
Description: New York: Riverhead Books, 2020. |
Includes bibliographical references and index.
Identifiers: LCCN 2019022493 (print) | LCCN 2019022494 (ebook) |
ISBN 9780735211605 (hardcover) | ISBN 9780735211629 (ebook)
Subjects: LCSH: Avery, John, active 1695. | Pirates— Great
Britain—Biography. | Piracy—Economic aspects—Great Britain—History.
| Piracy—Economic aspects—Mogul Empire—History. | International
economic relations—History. | Great Britain—Economic conditions—17th
century. | Mogul Empire— Economic conditions—17th century. | Mogul
Empire—Foreign economic relations—Great Britain. | Great
Britain—Foreign economic relations—Mogul Empire.
Classification: LCC G537.A9 J64 2020 (print) | LCC G537.A9 (ebook) |
DDC 910.4/5—dc23
LC record available at https:// lccn.loc.gov/2019022493
LC ebook record available at https:// lccn.loc.gov/2019022494

First Riverhead hardcover edition: May 2020
First Riverhead trade paperback edition: May 2021
Riverhead trade paperback ISBN: 9780735211612

Printed in the United States of America
3rd Printing

Book design by Amanda Dewey
Map by Jeffrey L. Ward

For Alexa

CONTENTS

Introduction 1

I. THE EXPEDITION *11*

1. Origin Stories *13*
2. The Uses of Terror *19*
3. The Rise of the Mughals *30*
4. *Hostis humani generis* *39*
5. Two Kinds of Treasure *45*
6. Spanish Expedition Shipping *55*
7. The Universe Conqueror *62*
8. Holding Patterns *68*

II. THE MUTINY *77*

9. The Drunken Boatswain *79*
10. The *Fancy* *88*
11. The Pirate Verses *96*
12. Does Sir Josiah Sell or Buy? *102*
13. West Wind Drift *111*
14. The *Ganj-i-Sawai* *120*
15. The *Amity* Returns *125*

16. She Fears Not Who Follows Her *130*

17. The Princess *135*

III. THE HEIST *141*

18. The *Fath Mahmamadi* *143*

19. Exceeding Treasure *150*

20. The Counternarrative *157*

21. Vengeance *164*

22. A Company at War *170*

IV. THE CHASE *181*

23. The Getaway *183*

24. Manifest Rebellion *189*

25. Supposition Is Not Proof *193*

26. The Saltwater *Faujdar* *198*

27. Homecomings *203*

V. THE TRIAL *209*

28. A Nation of Pirates *211*

29. The Ghost Trial *217*

30. What Is Consent? *224*

31. Execution Dock *238*

Epilogue: Libertalia *243*

ACKNOWLEDGMENTS   *257*

NOTES   *261*

BIBLIOGRAPHY   *271*

INDEX   *277*

Elegant and excellent was the pirate's answer to the great Macedonian Alexander, who had taken him: the king asking him how he dare molest the seas so, he replied with a free spirit, "How dare thou molest the whole world? But because I do with a little ship only, I am called a thief: thou doing it with a great navy, art called an emperor."

—St. Augustine, *The City of God*

Suffer pirates and the commerce of the world must cease.

—Henry Newton

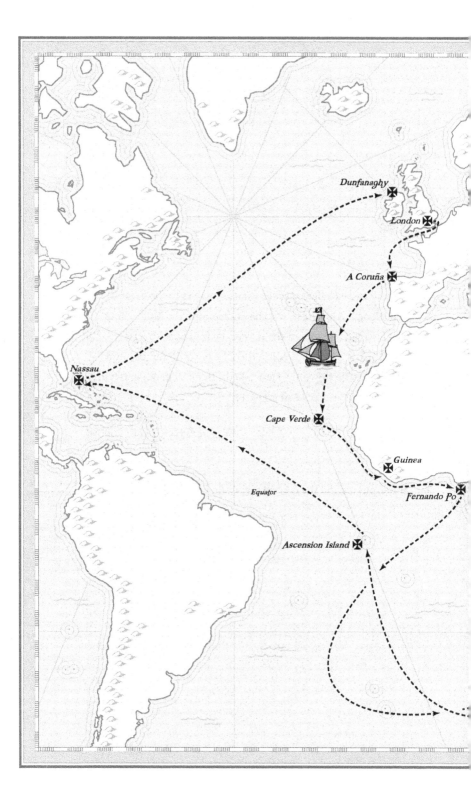

THE ROUTE
of
CAPTAIN EVERY
and
THE FANCY
1693–1696

Attack on
the Gunsway

• Surat

• Bombay

Perim ✠

Mayd ✠

Equator

Comoros Islands ✠

Augustin Bay ✠

✠ Reunion

© 2020 Jeffrey L. Ward

The Indian Ocean, west of Surat

September 11, 1695

⸺⸺

On a clear day, the lookout perched atop the forty-foot mainmast of the Mughal treasure ship can see almost ten miles before hitting the visual limits of the horizon line. But it is late summer, in the tropical waters of the Indian Ocean; the humidity lingering in the air draws a hazy curtain across the spyglass lens. And so by the time the English vessel comes into focus, she is only five miles away.

The existence of an English ship in these waters is hardly noteworthy. They are only a few days' sail from Surat, one of India's most prosperous port cities, and the original headquarters of the East India Company. At first sight, the lookout doesn't even think it necessary to sound an alarm. Yet as the seconds pass, as the blurred shape of the boat looms in the spyglass, something catches his attention in the approaching vessel: not her colors, but her speed in the water. The ship is in full sail, he can see now, running before the wind. And she is moving fast, at least ten knots, maybe more—easily twice the top speed of the treasure ship. The lookout has never seen a ship sail with such velocity across the open water.

By the time the lookout alerts the crew below him, the English ship is already visible to the naked eye.

From his vantage point on the quarterdeck, the captain of the Indian ship still has little reason to fear the approaching vessel, however fast she might be. He has eighty cannons lining his gun decks, supported by four hundred muskets and nearly a thousand men. From what he can make out, the English ship cannot have more than fifty cannons and a fraction of his crew. Even if she is under the command of pirates on the attack, the captain has been at sea for months without incident; he has sailed unchallenged through the notorious pirate's nest at the mouth of the Red Sea. Now he is practically in sight of his home port in Surat. What pirate would dare to challenge him in these waters, with so little firepower?

But the captain does not know the long history that has brought these two ships together. He does not know that the men aboard the English vessel have traveled thousands of miles to get this close to a ship returning to harbor with unimaginable riches in its hold, that they have waited more than a year for this precise opportunity. He does not know what these men are capable of, the crimes they have already committed.

And he does not know the near future, the two improbable events that are about to unfold within seconds of each other, radically undermining his advantage.

The sequence begins with the smallest of mistakes. An inexperienced gunner packs an extra ounce or two of gunpowder into the chamber of a cannon. Or perhaps, days or weeks earlier, the gun crew fails to clean the cannon properly, and a residue of gunpowder remains in the chamber, unnoticed. Or perhaps the chain of events starts much further back, in a blast furnace somewhere in India, where a minuscule flaw is formed in the cast-iron "reinforce" that houses the ignition chamber, a flaw that goes undetected for years, slowly weakening with each blast, until one day it fails.

A cannon is a simple piece of technology. It is a device that takes the multidirectional energy of an explosion and channels it down a single pathway, defined by the cannon's bore, the long cylinder down which the cannonball travels on the way to its intended target. The physics of the device are so intuitive that cannons were almost immediately invented once human beings first hit upon the chemical cocktail of sulfur, charcoal, and potassium nitrate—what we now call gunpowder—roughly a thousand years ago. Other uses of explosions—internal combustion engines, hand grenades, hydrogen bombs—would take centuries to emerge. But the cannon seems to have been a logical corollary to the discovery of gunpowder's propulsive force. Almost as soon as we had figured out how to make an explosion, we figured out how to harness that energy to shoot a heavy projectile through the air at high speeds.

The cannon was simple and effective enough that its basic design and operation went unchanged for hundreds of years. Gunpowder is funneled down the bore to the ignition chamber in a process called muzzle-loading, and then packed by a wadding conventionally made of paper. A cannonball is then inserted and wedged up against the wadding. Above the chamber, a small tube leads up to the top of the cannon, ending in a tiny opening called the touch hole. A fuse trails down from the touch hole into the chamber. In normal operations, a gunner lights the fuse, and a moment later, the gunpowder ignites, triggering a devastating blast of energy contained on all but one side by the sturdy bonds of iron and carbon in the cast-iron reinforce. Almost all the energy of the explosion rushes out the bore, propelling the cannonball with it.

Yet for all its strength, the crystalline structure of cast iron can sometimes be marred by invisible impurities, particularly when the ratio of carbon to iron is not properly regulated, making it prone to catastrophic failure. And when the cast-iron reinforce that houses the ignition chamber fails, a cannon ceases to be a cannon. It becomes a bomb.

The four members of the gun crew standing next to that cannon, as its reinforce shatters into hundreds of fragments, are dead before the sound of the explosion reaches their ears. And they are the lucky ones. The gunpowder ignition creates a sudden increase in air pressure inside the chamber; air that is normally fifteen pounds per square inch in a matter of milliseconds is pressurized to more than a thousand pounds per square inch. A blast wave surges out in all directions, traveling at more than twenty thousand feet per second, ten times the speed of sound. The wave severs arms and legs from the torsos of the nearby gunners, ruptures their organs; the combination of heat and pressure liquefies their eyeballs. By the time the second wave of energy—called the shock wave—rolls in, carrying the cast-iron fragments of the cannon at supersonic speeds, the gun crew has ceased to exist. The shock waves and the cannon fragments simulate the effects of a modern-day nail bomb, burrowing into the flesh and bones of other gun crews standing farther away. Hands and ears and legs are blown off; vital organs are punctured. Within a matter of seconds, the gun deck of the ship is coated with blood and tissue. And then, as the blast wind rushes back to fill the vacuum caused by the explosion, the wooden planks of the ship catch fire.

A FEW HUNDRED FEET AWAY, on the gun deck of the English ship, powder boys shuttle ordnance to the crews gathered around their cannons. There is a mechanical, rhythmic order beneath the surface chaos of the scene. An officer surveys a collection of gun crews, barking commands at regular intervals. "Cast loose your gun," he shouts, and the crews free the cannons from the tackles that hold them lashed to the ports. "Level your gun . . . Out tompion . . . Run out your gun . . ." With each order, the crews heave and thrust and balance in synchronized patterns, the deadly ballet of a ship engaged in battle. The midshipman commands, "Prime!" and a thin line of powder is poured down each touch hole. The crews do not notice the explosion that has unleashed such devastation on the gun deck of the Indian ship; they are intent on the most challenging of the midshipman's commands, the final one before the firing: "Point your gun."

Aiming a cannon on a seventeenth-century vessel is not even an art, much less a science. Calculating the proper trajectories for a projectile hurtling through the air at hundreds of miles an hour is hard enough on land. In fact, few practical tasks have a longer history of inspiring mathematical ingenuity than figuring out the trajectory of a projectile. Some of the first differential equations were developed to predict the flight of a cannonball shot. Many of the original computers built during World War II were explicitly designed to calculate rocket trajectories. But standing there in the middle of the Indian Ocean in 1695, heaving a five hundred pound cannon toward a small opening as the ship rolls with the waves, the idea of taking true aim is almost laughable. There is no time for math. You point the gun in the general direction of the opposing vessel; you listen for the officer's command; you hope for the best.

Despite the inexact nature of these weapons, every now and then the physics comes together to produce the perfect strike. Just minutes after the explosion on the Indian ship, one of the cannonballs released in the broadside hurtles across the gap that separates the ships and collides directly with the base of the mainmast of the Indian vessel: the single most devastating blow you can deliver with one cannonball. The mast collapses, and a tangled heap of rope and canvas crashes onto the deck. Without the square rig sails, the Indian ship cannot harness the energy of the wind with anywhere near the efficiency that it possessed seconds before. Still ablaze and bloodied from the cannon explosion, the treasure ship is suddenly defenseless. Within minutes, the British have boarded her.

WHAT ARE THE ODDS of these two events happening in the same instant? Explosions had plagued cannon design from the very origins of the technology, and they remained an issue well into the modern age. (An 1844 explosion during a cannon demonstration killed the US Secretary of the Navy and the Secretary of State, and very nearly killed President John Tyler.) But the likelihood of a cannon exploding on any given firing was very small, probably less than one in five hundred. The odds of hitting the lower half of the mainmast with a single shot were not much better: The beam of the mast is two feet wide on a ship that is more than two hundred feet long. Aim too low, and the broadside lands in the water or barrels into the gun decks. It is at least a one-in-a-hundred shot. Thanks to the laws of probability being devised by Blaise Pascal right around the same period, we know that the odds of two unrelated events happening simultaneously can be calculated by multiplying the individual odds of each event. If you could somehow replay those seconds five thousand times, the combination of the cannon explosion and the mainmast strike might never happen again.

You can measure the difference between these two events happening and not happening in inches and ounces. Remove that tiny impurity from the cast-iron reinforce, or shift the cannon an inch to the left as it fires, and the Indian vessel easily shakes off its underpowered attacker. But like the cannon explosion itself, an almost imperceptible difference—a few ounces of extra gunpowder—can trigger nonlinear results. In the case of these two ships confronting each other in the Indian Ocean, those nearly microscopic causes will trigger a wave of effects that resonate around the world. Most confrontations like this one, viewed from the wide angle of history, are minor disputes, sparks that quickly die out. But every now and then, someone strikes a match that lights up the whole planet. This is the story of one of those strikes.

You can think of the narrative that follows as a kind of hourglass. At the waist of the hourglass—its center point—are those few seconds on the Indian Ocean in 1695: the exploding cannon, the destroyed mainmast. Before the waist is the layered history of events that made those extraordinary seconds possible. After the waist is the sweeping—truly global—chain of events those seconds unleashed.

To do justice to that story—particularly on the pre-history side of the hourglass—the chains of cause and effect have to run at different scales. Some events have short-term causes, like that exploding cannon or the direct hit on the mainmast. But other causes roll in more slowly, like those that ultimately brought so much treasure aboard the Mughal ship, or the ones that ultimately incentivized a small band of human beings to deliberately become pirates. A full account of the events demands that you break out of the boundaries of period histories and traditional biographies. You have to jump around in time to get the facts right. Linear chronology makes for good popular storytelling, but it doesn't always capture the deep causes that drive history. Some causes are proximate, in the moment. Some are echoes of distant shock waves, still reverberating a hundred—or a thousand—years later.

AT ITS SIMPLEST, this is the story of a rogue pirate and his sensational crime. While piracy is an ancient profession, the most famous pirates of history would not take the stage until two decades or so after the events chronicled in this book. But that "golden age" generation—Blackbeard, Samuel Bellamy, Calico Jack—was very much inspired by the notorious acts described here, and the legends that spun up around them. While the pirate at the center of this story is not as famous today as those iconic figures from the golden age, he had a more significant impact on the course of world events than Blackbeard and his peers. This book is an attempt to measure that impact, to chart its boundaries. It tells the story of the individual lives caught up in the crisis that erupted after the events of September 1695, but also the stories of a different kind of character, one step up the chain: forms of social organization, institutions, new media platforms. One of those institutions was as ancient as piracy itself: the autocratic theocracy of the Mughal dynasty. The others were just coming into being: the multinational corporation, the popular press, the administrative empire that would come to dominate India starting in the middle of the next century. In part, this is a book about a deeply flawed man who became a pirate for a very short, but very eventful period of his life, a life that is more intriguing and more mysterious the further you dig into it. But it also describes a different kind of life span, the story of how some of the most powerful institutions of modern history went from a fledgling idea, promising but not inevitable, to world conquest.

This book is not intended to be anything close to a comprehensive account of the rise of those institutions. It focuses, instead, on one key challenge that threatened to undermine their ultimate triumph. We tend to think of grand organizations like corporations or empires coming about through deliberate human planning: designing the conceptual architecture for each imposing structure, brick by brick. But the shape an institution ultimately takes is not so much designed in advance by a master engineer as it is carved away by challenges to its outer boundaries, the way a coastline is partly formed by an endless battering of much smaller waves. The core values of long-standing institutions are often first established by the founders and the visionaries that traditional histories foreground, for understandable reasons. But the ultimate structure of those organizations—the limits of their power, the channels through which they can express that power—are more often than not defined by edge cases, by collisions at their borders, both geographic and conceptual.

Sometimes those collisions involve equally powerful organizations—as in the clashes between the Mughal empire and the British Crown that animate so many of the events in this book. But sometimes the collision comes via a much smaller force: a ship sailing the Indian Ocean with less than two hundred men on board, led by a captain who has been dreaming of this encounter for almost two years.

The crew had christened the ship the *Fancy* fourteen months before that showdown in September 1695. Their captain, on the other hand, went by many names.

Part One

THE
EXPEDITION

ORIGIN STORIES

Newton Ferrers, Devonshire

August 20, 1659

Sometime around the year 1670, a young man from Devon in the West Country of England joined the Royal Navy. Given that he would spend the rest of his adult life on the water, it is possible that he willingly volunteered for service. There were economic advantages to volunteering: the navy offered two months' salary in advance, though it was expected that the new recruit would spend some of those funds purchasing equipment (including the hammock they would sleep in on board). New volunteers were also protected from creditors if they owed less than £20. But roughly half the sailors in the Royal Navy had been forced into service thanks to one of the most notorious institutions of the period: the impress service.

To be a young man in England in the seventeenth century—particularly a young man of limited means—was to live with a constant background fear of the impress service, roving bands of informal agents for the Royal Navy known colloquially as "press-gangs." Impressment was a kind of hybrid of the modern military draft and state-sponsored kidnapping. A seventeen-year-old could be standing on a street corner, minding his own business, and out of nowhere a press-gang could swoop in and make him a *Godfather*-style offer he couldn't refuse: he could voluntarily join the navy, or he could be forced into service under worse terms. The choice was his to make—as long as it ended up with him on a Royal Navy ship.

Newly impressed sailors confronted a grim reality once they had been loaded onto the guard ships where the men were held until they could be assigned to a specific ship. An eighteenth-century tract called *The Sailors Advocate* described the scene: "They found seldom less aboard the Guard-ship, than six, seven, or eight hundred at a time in the same condition that they were in, without common conveniences, being all forced to lie between decks, confined as before, and to eat what they could get, having seldom victuals enough dressed, which occasioned distempers, that sometimes six, eight, and ten, died a day; and some were drowned in attempting their escape, by swimming from the Guard-ship; many of whose bodies were seen floating upon the River. . . ."

Impressment arose in part because the age of exploration created a demand for labor at sea that could not be met through normal financial incentives. But it also arose because of changes on land. The shift from late feudalism to early agrarian capitalism, the great disruption that would fuel the growth of the metropolitan centers in the coming centuries, had disgorged a whole class of society—small, commons-based cottage laborers—and turned them into itinerant free agents. By the late 1500s, the explosion of vagabonds made them public enemy number one, triggering one of the first true moral panics of the post-Gutenberg era. Everywhere there were wanderers, whole families lost in the changing economic landscape. Serfs once grounded in a coherent, if oppressive, feudal system found themselves flotsam on the twisting stream of early capitalism. To everyone sitting on the banks above that stream, the change must have seemed something like the modern fantasies of zombie invasions: you wake up one day and realize that the streets are filled with people who not only lack homes, but also suffer from some other, more existential form of homelessness—not even knowing what *kind* of home they should be seeking.

In 1597 Parliament passed a vagrancy act that attempted to combat the scourge of homelessness. The language of the act includes an almost comical catalog of the various species of vagabonds currently at large on the public roads and in the town squares of England:

> Wandering scholars seeking alms; shipwrecked seamen, idle persons using subtle craft in games or in fortune telling; pretended proctors, procurers, or gatherers of alms for institutions; fencers, bear wards, common players, or minstrels; jugglers, tinkers, peddlers and petty chapmen; able-bodied wandering persons and laborers refusing to work for current rate of wages; discharged pensioners; wanderers pretending losses by fire; Egyptians or gypsies.

The Vagabond Act had a clear message to local authorities: any of these characters were to be "stripped naked from the middle upwards and openly whipped until his or her body be bloody, and then passed to his or her birthplace or last residence." But the act also empowered the press-gangs. If the wandering scholars and jugglers didn't want to be stripped naked and openly whipped, they could always join the Royal Navy. What better way to clear the streets of the refugees from a fallen feudal order than to send them off to sea?

Whether he joined the Royal Navy on his own accord or was forced into service by the press-gangs, the Devonshire sailor would have grown up in a culture that was heavily shaped by stories of seafaring life. No region of Britain is more closely associated with maritime adventure than the West Country, the rugged moorlands that jut out into the Atlantic, wedged between the English and Bristol Channels. Almost all the legendary sea dogs of the Elizabethan age hailed from the region. Both Walter Raleigh and Francis Drake were born in Devon. While the West Country mariners led many naval battles on behalf of the Crown—including the sinking of the Spanish Armada in 1588—many of them also crossed over into piracy. (The two most notorious pirates of the 1700s—"Black Sam" Bellamy and Blackbeard—were also West Country natives.) The prominence of the swashbuckling lifestyle had geological roots: the West Country's position at the mouth of the English Channel gave its captains unrivaled access to the shipping networks of Europe, and the many coves and inlets carved into the coastline made the landscape ideal for smugglers. The link between piracy and Devonshire lives on in our speech patterns more than three hundred years after that Devonshire boy first joined the navy. When we adopt a stereotypical pirate accent today—"Arr, shiver me timbers"—we are, unconsciously, mimicking the lilt and idiosyncratic grammar of West Country–vernacular English.

The mystery that surrounds the life of the Devonshire sailor begins with his name. The first biographical account of his exploits, published in 1709, referred to him as Captain John Avery. As a young man, he seems to have briefly adopted the alias of Benjamin Bridgeman, though his nickname, "Long Ben," has led some historians to speculate that Bridgeman was his original name and Avery the alias. Most scholars agree that he was born near Plymouth, in Devonshire, on the southwest coast of England. An acquaintance would testify under oath in 1696 that the sailor was a man of about forty years of age, dating his birth back to the late 1650s. Parish records in Newton Ferrers, a village on the River Yealm southeast of Plymouth, note the birth of a child to John and Anne Avery on August 20, 1659. Perhaps that child grew up to be the notorious Henry Avery, the most wanted criminal on earth. Or perhaps the real Avery was born in another West Country village in that same period. In part because a family by the name of Every had been prominent landowners in Devonshire for centuries before his birth, many accounts of his life refer to him as Henry Every. Almost every legal document written in English that would eventually mention his name spelled it "Every," and the one piece of his correspondence that survives was signed "Henry Every." Every was the name most often invoked by the public after he became one of the most notorious men in the world. For that reason alone, it seems appropriate to call him Henry Every.

Almost nothing is known about Henry Every's childhood. A memoir published in 1720 keeps his early years heavily veiled: "In the present Account, I have taken no Notice of my Birth, Infancy, Youth, or any of that Part; which, as it was the most useless Part of my Years to myself so 'tis the most useless to any one that shall read this Work to know, being altogether barren of any Thing remarkable in it self, or instructing to others." Given that this memoir was almost certainly a sham—some believe it was, in fact, the work of Daniel Defoe—the omission of childhood details most likely reflects how barren the historical record was, and not the uselessness of Every's actual upbringing.

No doubt young Henry Every (or Avery or Bridgeman) grew up hearing folk tales about the globetrotting exploits of Drake and Raleigh, both of whom skirted the line that separated pirate from privateer in their careers at sea. (As we will see, the legal conventions of the period kept that line deliberately blurry.) The faux memoirs claim that his father had served in the Royal Navy as a trading captain; the Devonshire Every clan included at least a few captains in their family tree. Whatever the details, Every seems to have been, as he puts it in the fictional memoirs, "bred to the Sea from a Youth." Appropriately enough, the first real biographical detail we have of Every's life—beyond those parish records in Newton Ferrers—is that he joined the Royal Navy, likely as a teenager.

The fog around the birth of that Devonshire sailor is almost as thick as the one that surrounds his death. The truth is we don't really know when or where he was born, or even what his name actually was. It is fitting that there should be a certain blurriness to Henry Every's roots. All the great legends have palimpsest narratives of their origins, different plots layered and threaded together through rumor and hearsay and the subtle transformations that befall any story passed down from generation to generation. For a time, Henry Every was a legend as widely known as any in the pantheon, a hero and inspiration to some, a ruthless killer to others. He was a mutineer, a working class hero, an enemy of the state, and a pirate king.

And then he became a ghost.

2

THE USES OF TERROR

The Nile Delta

1179 BCE

To modern eyes, the hieroglyphs that line the external north-west wall of Medinet Habu, the Mortuary Temple of Ramses III, are inscrutable, written in a language that only a small group of Egyptologists can now read. But the images etched in bas-relief on the temple walls are easily deciphered. They depict a scene of terrible carnage: warriors carrying javelins and daggers, fortified by shields and Aegean armor, fending off a shower of arrows; an officer wearing Egyptian headgear a split second away from decapitating a fallen enemy; a bloodied mound of corpses signaling the total annihilation of the invading forces. The images—and the hieroglyphs beside them—tell the story of one of the ancient world's largest naval battles, the clash between Egyptian forces and a band of itinerant raiders known today as the Sea Peoples. Because they left behind archaeological wonders like the temple of Ramses III and the pyramids, not to mention the treasures of Tutankhamun, the Egyptian dynasties to which Ramses III belonged have long held a vivid place in our historical imagination. Every grade-schooler can tell you something about the pharaohs. The Sea Peoples did not attract the same legacy, largely because they spent most of their prime living an entirely nautical existence. They did not leave temples or monuments behind to astound tourists three thousand years after their demise. They did not pioneer new forms of agriculture, or compose philosophical tracts. They left no written records at all. But the Sea Peoples should loom larger in the modern memory of the ancient world for one reason. They were the first pirates.

The geographic origins of the Sea Peoples remain a matter of debate among historians. The prevailing theory is that the Sea Peoples were a collection of refugees from Mycenaean Greece who first took shape as a coherent cultural group at the end of the Bronze Age. Some of them were warriors and mercenaries, others ordinary laborers who had previously been employed at borderline slave wages building the immense infrastructure and fortifications that marked the heyday of the Mycenaean age: the network of roads in the Peloponnese or the deepwater harbor at Pylos. Their origins are necessarily murky because the Sea Peoples ultimately became, like so many pirate communities since, a multiethnic group, defined not by their allegiance to a single city-state or emperor, but rather by their own elective allegiance to the floating community they had formed. Their homeland was the Mediterranean, and the ships they sailed upon it. They built customs and codes that helped define their tribal identity: they sported distinctive horned helmets—clearly visible in the Ramses III engravings—and their ships were adorned with figureheads of birds. But what made them so unusual was their rootlessness, both in the sense of leaving behind their geographic homelands and of being perpetually in motion, never stopping long enough to put down roots.

That rootlessness implied a political stance, one that would be adopted by the most radical of pirates in the centuries to come. The

Sea Peoples did not respect the authority of the existing land-based regimes that surrounded the Mediterranean. They were not bound by the laws of terrestrial states. This is one of the key ways in which the Sea Peoples mark the point of origin for piracy as a form of self-identity. Before the Sea Peoples, there were no doubt acts of piracy committed on the open sea; as soon as human beings began transporting valuable goods via ship, you can be sure there were criminals scheming to intercept those vessels and run off with the loot. But a true pirate is not just a subclass of criminal like a bank robber or a petty thief. Most people we consider to be criminals are people who break the law deliberately, but who still, in other aspects of their lives, acknowledge the rule of law. They get driver's licenses, and pay taxes, and vote. They consider themselves citizens, just not entirely law-abiding ones. To be a true pirate implies a broader disavowal. The pirate renounces the long-distance authorities of nations and empires. This is why the pirate flags that every grade-schooler can recognize today—centuries after they were last flown in earnest—carried so much symbolic heft. The pirate sails beneath the colors of his or her own rogue state, "reckless wanderers of the sea," as Homer described them in *The Odyssey*, "who live to prey on other men."

Not all pirates were willing to make such a complete break with their national allegiances, of course. (The tension between open rebellion and patriotic loyalty would shape many of the events in Henry Every's brief career as a pirate.) But the pirates' willingness to challenge the legal and geographic boundaries of state power—not to mention their fondness for pillaging—made them frequent enemies of centralized authority. Nimble, unburdened by legal or moral restraints or by state bureaucracies, the pirates had many advantages over their larger antagonists. But they were not invulnerable to a concerted effort by a centralized government to defeat them. In 1179 BCE, the Sea Peoples launched an attack on Ramses's forces in the

Nile Delta. Anticipating their attack, the pharaoh had constructed ships designed specifically to combat the Sea Peoples' naval advantage. He set up a network of scouts to watch for invading ships, and anchored his new fleet just out of sight in the many channels feeding the delta. The drawings at Medinet Habu show the Sea Peoples without oars in their galleys, suggesting that they were ambushed. The scenes bring to mind the storming of the beaches at Normandy: a scattered mass of boats washing ashore and men scrambling off into the waves, only to be picked off by distant Egyptian archers. Many bled to death in the shallow water.

For once, it was the Sea Peoples' turn to feel the wrath of a merciless military force. "They were dragged, overturned, and laid low upon the beach; slain and made heaps from stern to bow of their galleys, while all their things were cast upon the water," Ramses III inscribed on the walls of Medinet Habu. "His majesty is gone forth like a whirlwind against them, fighting on the battle field like a runner," other hieroglyphs at his tomb attest. "The dread of him and the terror of him have entered in their bodies; [they are] capsized and overwhelmed in their places. Their hearts are taken away; their soul is flown away."

The inscription was more prophetic that its authors might have realized at the time. After their defeat at the Nile Delta, the Sea Peoples exited almost immediately from the world historical stage. Scholars are as divided about their ultimate fate as they are about their enigmatic roots. The ones who were not executed after the Battle of the Delta appear to have been scattered along the eastern frontier of the Egyptian realm, some of them on the Palestinian coast. But as a coherent—if itinerant—group, they had ceased to exist by the time Ramses died in an apparent assassination in 1155 BCE. In this respect, too, the Sea Peoples established a tradition that many a pirate would emulate in the centuries to come. Some pirates go out in a blaze of glory. Some end up hanging on a gallows. But some of them just disappear.

The legacy of the Sea Peoples also included another key element that would come to define the pirate culture of Every's time: the tactical deployment of spectacular violence and terror. Under siege by the Sea Peoples, King Ammurapi of Ugarit—part of modern-day Syria—sent a desperate missive to another ruler in Cyprus: "My cities were burned, and [the Sea Peoples] did evil things in my country. . . . The seven ships of the enemy that came here inflicted much damage upon us." The inscription at the temple of Ramses III sounds a comparable note in its description of the coastal raids of the Sea Peoples: "All at once the lands were removed and scattered in the fray . . . A camp was set up in one place in Amor. They desolated its people, and its land was like that which has never come into being."

The carnage unleashed by the Sea Peoples was so extreme during their heyday in the thirteenth and twelfth centuries BCE that it provoked a massive crisis among the previously flourishing Mediterranean civilizations of the Bronze Age. Today, this period is known as the Late Bronze Age Collapse, one of those stretches of history where the march of technological progress reverses course. The great palace societies of Greece and the Levant disintegrated into loosely organized village cultures, after the Sea Peoples laid waste to their coastal capitals. They brought a near apocalyptic destructiveness to their interactions with all land-based communities, a violence that seemed almost arbitrary in its intensity. The Sea Peoples did not invade lands to claim them as their own, or extract treasure or slaves to bring back to their homeland. They torched the great capitals of the Bronze Age just to watch them burn. While they lacked the armies and fortresses of their mainland foes, their strategic use of terror allowed them to wage what we now call "asymmetric" warfare: a much smaller force successfully challenging a much larger one.

From its beginnings, piracy has shared many key traits with the modern concept of terrorism, both in terms of its hold on the popular imagination and in its legal definition. One of the first known uses of the word "terrorism" in the English language appears in a letter written to Thomas Jefferson in 1795 by James Monroe, then the American ambassador to France. Writing from Paris a year after the execution of Robespierre, Monroe referred to the Jacobin attempt to restore "terrorism and not royalty." The terminology appears to have spread quickly among the American political elite. In a letter written just a few weeks after Monroe's, John Quincy Adams alluded to the "partisans of Robespierre's dominion" as "terrorists."

The sense of terrorism as a tool to advance radical political values through the use of targeted public violence belongs to both its original use and its modern deployments. But in one crucial sense, the contemporary definition no longer matches its original meaning. Until the twentieth century, the idea of terrorism took its cues from the actions of the the Comité de Salut Public (Committee of Public Safety) and other arms of the French revolutionary government. Terror, in other words, was a political tactic that belonged to the state apparatus. It wasn't until the rise of the anarchists a century later that the idea of terrorism would come to be associated with non-state actors, small groups disrupting civilian life with carnage and explosions as a way of fighting a proxy war with massive governments and military powers. Robespierre's terror took the state's legal monopoly on violence to devastating extremes. It was a way of making a ruling power even more formidable. Modern terrorism does the reverse: it grants a disproportionate power to small bands of insurgents and shadow networks. The whole notion of asymmetric warfare that characterizes so many contemporary military conflicts—in which a superpower finds itself engaged in battle with an enemy thousands of times smaller in terms of manpower and military might—has its roots in this reversal of terrorism's meaning. Modern terror is a force multiplier. You don't need a vast standing army or a fleet of aircraft carriers to create a gnawing sense of fear among millions of people. You just need a few well-placed explosives—or even box cutters—and a network of media outlets willing to amplify news of your attack.

While the actual etymology of the word "terrorism" dates back to Robespierre's reign, some of the first real practitioners of terrorism's contemporary form—extreme violence carried out by non-state actors, creating disproportionate effects through media dissemination—were pirates. And the first convincing proof that such a strategy could work—that a handful of men could effectively hold entire nations hostage with a few acts of grotesque barbarity—would play out in the clash between the *Fancy* and the Mughal treasure ship in 1695.

There was precedent for that strategic terror, of course—starting with the legendary brutality of the Sea Peoples. Another pioneer in that bloody tradition was a French noblewoman named Jeanne-Louise de Belleville, born in the first year of the fourteenth century. In the middle of the Hundred Years' War between France and England, Belleville's second husband, Olivier de Clisson, was executed by the French king Philip VI for treason, his head mounted on a spike and publicly displayed in Nantes, Brittany, near Clisson's estate. Outraged by the king's actions, Jeanne sought vengeance by selling her lands and property and assembling a small fleet of three ships. For dramatic effect, she painted the ships black and hoisted sails that had been dyed blood red. Legend has it that she prowled the English Channel for thirteen years, assisted by two of her sons, attacking French ships and decapitating the supporters of Philip, always leaving a handful of survivors to bring back word to the mainland of the "Lioness of Brittany."

"Dead men tell no tales" is the pirate mantra often invoked as a justification for killing off one's enemies, but for pirates like Clisson and her descendants, the slogan has an alternate meaning: dead men by definition can't amplify the pirate's reputation for bloodlust and savagery if they're tossed overboard. By the so-called golden age of piracy, the generation of pirates that followed Henry Every, it had become standard practice to grant mercy to a few lucky survivors so they could return home with tales of terror on the seas. Living, as she was, in a pre-Gutenberg era, the Lioness of Brittany could only send messages that circulated through palace rumor and personal correspondence. But Every and his descendants had a vibrant media apparatus through which they could broadcast their atrocities: the pamphlets, newspapers, magazines, and books that shaped so much of popular opinion in European and Colonial American cities during that period. Many of the conventions that we associate with "tabloid" media—hastily written, often fabricated stories of sensational violence—were first developed to profit off the distant actions of men like Henry Every and the pirates who followed him in the early 1700s. If Every was, in his prime, the descendant of mythic seafaring men like Odysseus, he was also an augur of another kind of larger-than-life figure: the killer that captivates a nation with his outlandish crimes, like John Wayne Gacy, Son of Sam, Charles Manson.

We tend to think of the pamphleteers and early journalists of the Enlightenment as a refined intellectual class, filing witty copy for the *Tatler* from their coffeehouse off the Strand. But even in those formative early years of print media, there was no shortage of sensationalism. Enterprising publishers would hawk special-edition broadsides at public executions, promising morbid details about the original crime. Almost two centuries before Jack the Ripper became the first celebrity serial killer, the pamphleteers were already making quick money celebrating violent criminals. And no class of criminal captured the popular imagination more than the pirates.

The most extreme serial killer narratives of the modern age have nothing on the gruesome inventories of pirate torture published during this period. A French pirate by the name of François l'Ollonais was reported to have "ripped open one of the prisoners with his cutlass, tore the living heart out of his body, gnawed at it, and then hurled it in the face of one of the others." The *American Weekly Mercury*, an early colonial newspaper, relayed a particularly appalling account of the British pirate Edward Low: after a merchant captain allegedly tossed a satchel of gold overboard, Low "cutt off the said Masters lipps and broyl'd them before his face, and afterwards murder'd the whole crew being thirty two persons." In one later version of the story, worthy of a modern-day Hannibal Lecter novel, the maniacal pirate forces the captain to eat his own lips after broiling them.

No doubt many of these stories were exaggerated to sell copy. But accounts of pirate atrocities drew from the transcripts of legal trials. These publications—often printed within days of a delivered verdict—began a long tradition of media amplifying the reach of scandalous court cases. One of the most appalling in the genre was an account of a "Captain Jeane of Bristol," accused of torturing and murdering a teenaged cabin boy who had dared to steal a dram of rum from his quarters. The book was published under the title *Unparallel'd Cruelty*, a title that almost seemed understated given its account of the boy's agonizing and prolonged murder: strung up on the mainmast for nine days, whipped, forced to drink the captain's urine, among other atrocities.

The sadistic violence of Captain Jeane did not end well for the pirate. He was condemned to death and hanged in an unusually brutal fashion, dangling by the neck for eighteen minutes before dying.

But in many cases, the mythologies of pirate brutality were not just a symbol of their deranged mental state. If the pamphleteers of London or Boston had a financial incentive in tales of buccaneer brutality, so did the pirates themselves. By cultivating a reputation for bloodlust and mayhem, the pirates made their own jobs easier. A merchant captain who had just read an account of one of his peers being force-fed part of his own anatomy would naturally be more inclined to surrender his ship at the first sight of a black flag. There was, in other words, a method in the madness. In the economic historian Peter Leeson's study of the surprisingly rich economic systems that the pirates maintained—memorably titled *The Invisible Hook*— he describes the extremes of pirate violence as a kind of semiotic act:

> To prevent captives from withholding booty . . . pirates required a reputation for cruelty and barbarity. And adding madness to the piratical reputation didn't hurt either. Pirates institutionalized their reputation for ferocity and insanity into a piratical brand name through the same means Mercedes-Benz uses for this purpose: word of mouth and advertisement. Pirates didn't take out glossy ads in magazines. But they did make a point of publicizing their barbarity and madness so their reputation could strengthen and spread. What's more, pirates received advertisement for their reputation in popular eighteenth-century newspapers, which unwittingly contributed to pirates' ruthless brand name, indirectly facilitating pirates' profit.

While they were usually separated by thousands of miles of open ocean, the enterprising publishers of London, Amsterdam, and Boston were locked in a symbiotic embrace with the pirates themselves: the publishers needed stories of living hearts being torn out of chests to sell copy; the pirates needed those stories to circulate as widely as possible to instill fear in their prospective victims. The fact that the golden age of piracy coincides almost exactly with the emergence of print culture is no coincidence. Jeanne de Clisson may have made a name for herself in the fourteenth century by haunting the English Channel for a decade, but in general it was challenging to establish yourself as a pirate without the power of media amplification. If you wanted to make a living as a pirate, an appetite for cruelty and physical abuse was helpful. But it was even better to be famous.

3

THE RISE OF THE MUGHALS

The Bolan Pass
663 CE

With mountain peaks that rarely exceed ten thousand feet, the Central Brahui Range that runs through the center of modern-day Pakistan does not possess the same glamour as its neighboring range to the north, the Himalayas. But for many centuries, a fifty-five-mile stretch of valleys and gorges naturally carved through the limestone ridges of the Brahui served as the primary conduit linking the Arab world to the agricultural settlements of the Indus Valley and the wide sweep of the Indian subcontinent beyond them. Today, you can travel the Bolan Pass—named after the mountain stream that opened up the gateway through thousands of years of erosion—via car or rail. The pass was not always so accessible. A British military officer described the pass in a letter to the Royal Geographical Society in 1841: "Should there be rain in the higher part of the mountains, the stream at times comes down in an almost perpendicular volume, without warning, and sweeping all before it, as a friend of mine experienced, when he saw a party of men, horses, and camels, and all his property, borne down by it. . . . About thirty-seven men were washed away upon that occasion."

In 663 CE, only thirty-one years after the death of the prophet Muhammad, a Muslim military force successfully traversed the Bolan Pass and made their way down the foothills of the Brahui into the valleys of the subcontinent. (Among their ranks may have been a few religious disciples who had studied with Muhammad himself.) The raid marked the first time Islamic soldiers made contact with the Hindu cultures of India. At the time, the expedition through the pass seemed like the logical continuation of the furious three decades of conquest that had followed Muhammad's death. The birth of Islam is conventionally dated at 622 CE, with Muhammad's exodus from Mecca; by 650 CE, Muslim forces had toppled the last vestiges of the Roman Empire, seizing contemporary Syria, Egypt, Iraq, Iran, parts of North Africa, and most of Afghanistan. That Islam would continue its march into India seemed all but inevitable. Muslim traders had already started doing business in the port cities of western India, their merchant ships following the same path across the Arabian Sea that Henry Every's ship would take a thousand years later.

But the warriors who made it across the Bolan Pass in 663 would not prove to be conquerors. They were quickly rebuffed by a brahman by the name of Chach who ruled over the Sindh region during that period. But a half century later, Muhammad bin Qasim successfully returned to conquer Sindh and the Indus Valley. For the next few centuries, the lands shuffled back and forth between Islamic occupation and native rule, but the invaders never managed to control much of India beyond those northern regions. The Muslim interlopers came to be known as *mlecchas*: a dismissive term that suggested inferiority, and not a looming threat. In part, their conquest was limited by the natural barrier of the Thar Desert, which today defines the border between Pakistan and India. Trade, however, did manage to establish a constant web of interdependence between the two cultures. Islam created the first truly global integrated trading network in world history, reaching from western Africa all the way to Indonesia, but in that vast network, few trade routes were as lucrative as the one that brought Arabian horses to India in return for spices and cotton.

Global trade ultimately made India too wealthy for Islam's imperial ambitions to resist. From 1 CE to 1500 CE, no region in the world—including China—had a larger share of global GDP. Its copious supply of pearls, diamonds, ivory, ebony, and spices ensured that India ran what amounted to a thousand-year trade surplus. But no product ignited the imagination of the world—and emptied its pocketbooks—like the dyed cotton fabrics that would play such a critical role in the history of India. The link between cotton and the subcontinent is an ancient one. Archaeological excavations along the Indus River in modern-day Pakistan uncovered a few threads of dyed and woven cotton that had been affixed to a silver vase. The fabric is believed to have been created sometime around 2300 BCE, making it one of the earliest known examples of processed cotton fibers anywhere in the world. Herodotus took note of wild trees in India "which produce a kind of wool better than sheep's wool in beauty and quality, which the Indians use for making their clothes." From the beginning, cotton inspired technological innovation. The frescos in the legendary Ajanta Caves, dating back to roughly the same period, feature Indians working single-roller machines designed to extract the seeds from the cotton fibers, an early antecedent of Eli Whitney's cotton gin.

But the innovation that would most transform the subcontinent—and its economic relationship to the rest of the world—did not involve separating the seeds from their fibers; every society that domesticated cotton for textile use ultimately developed some kind of mechanical gin. What made Indian cotton unique was not the threads themselves, but rather their color. Making cotton fiber receptive to vibrant dyes like madder, henna, or turmeric was less a matter of inventing mechanical contraptions as it was dreaming up chemistry experiments. The waxy cellulose of the cotton fiber naturally repels vegetable dyes. (Only the deep blue of indigo—which itself takes its name from the Indus Valley where it was first employed as a dye—affixes itself to cotton without additional catalysts.) The process of transforming cotton into a fabric that can be dyed with shades other than indigo is known as "animalizing" the fiber, presumably because so much of it involves excretions from ordinary farm animals. First, dyers would bleach the fiber with sour milk; then they attacked it with a range of protein-heavy substances: goat urine, camel dung, blood. Metallic salts were then combined with the dyes to create a mordant that permeated the core of the fiber. The result was a fabric that could both display brilliant patterns of color and retain that color after multiple washings.

It is not known when this technique was invented. Almost certainly it was not discovered by a single inspired dyer, but rather it evolved over centuries of experimentation. By 327 BCE, when Alexander the Great launched his campaign into the subcontinent, the dyed cotton fabrics were so conspicuous that several of his generals featured them prominently in their accounts of the campaign. "There were in India trees bearing, as it were, flocks or bunches of wool," the Greek historian Strabo recounted, quoting the generals. "The linen made by them from this substance was finer and whiter than any other . . . The country produces colours of great beauty."

As Alexander's forces returned from India with word of this miraculous fabric, they helped initiate an obsession with Indian cotton that would ultimately wrap itself around the globe. That obsession arose from the confluence of three properties: the fabrics were soft; they could be dyed with vibrant patterns; and those patterns could be washed without losing color. No fabric in human history had combined those properties into a single cloth. Over the two millennia that separate Alexander's invasion and the battle between the *Fancy* and the treasure fleet, many fortunes were made unearthing and trading rare metals, or growing and selling valuable foodstuffs like sugar and pepper. But no product of art and manufacture during that period generated as much profit as the dyed cotton fabrics of India.

While India was a defining force in global trade from Roman times all the way into the age of exploration, India itself had only a marginal role in moving its product around the world. The historian Strabo recorded that 120 Roman ships a year, manned by Egyptian Greeks, would sail to India's southwest coast to trade silver and gold for cotton, jewelry, and spices. By the end of the millennium, that shipping network would be run almost exclusively by Muslim traders. The result was a geoeconomic system in which an artisanal Hindu society produced valuable goods, while surrounded by a membrane of Islamic merchants and sailors concentrated in the harbor cities that allowed those goods to circulate on the world market.

The question of why India itself never developed its own trade networks leads to one of the great "what if" thought experiments of global history. Had the subcontinent's combination of immense natural resources and technical ingenuity been matched with an equivalent appetite for seafaring trade, it is not hard to imagine India following the path to industrialization and global dominance *before* England made its great leap forward economically in the 1700s. One explanation for India's reluctance to trade lies in the Hindu prohibition against oceanic travel. According to the Baudhayana sutra, anyone "making voyages by sea" would lose their status in the caste system, a punishment that could only be absolved through an elaborate form of penance: "They shall eat every fourth mealtime a little food, bathe at the time of the three libations (morning, noon and evening), passing the day standing and the night sitting. After the lapse of three years, they throw off their guilt." The prohibition itself took only a few lines to spell out, but it cast a long shadow.

Some historians have argued that—prohibitions notwithstanding—India in the first centuries of the Common Era had more nautical expertise than the conventional historical account would have it. But for whatever reason, by the end of the first millennium CE, Muslim trading fleets had swooped in to dominate the flow of goods in and out of the subcontinent. Islam, in those early years, was as extroverted in its attitude toward commerce as India was introverted. Muhammad had been a trader, and his disciples recognized early on that selling people much-coveted products was a particularly effective way to start relationships that ultimately led to religious conversion. (The map of modern-day Islam is defined almost entirely by regions of the world where its traders did business a thousand years ago; most of Islam's military conquests from the period rejected the religion when the occupying armies left.) Of all the world religions circa 1000 CE, Islam was by far the most cosmopolitan; the most open to new encounters, often facilitated by commerce, with other cultures and religious traditions. They found the insular culture that they interacted with in those port cities to be baffling. "The Hindus believe that there is no country but theirs, no nation like theirs, no king like theirs, no religion like theirs, no science like theirs," the Islamic scholar Al-Biruni noted in the eleventh century. "Their haughtiness is such that, if you tell them of any science or scholar in Khurasan or Persia, they will think you to be both an ignoramus and a liar. If they travelled and mixed with other nations, they would soon change their minds."

Despite their differences, the Hindu and Muslim cultures maintained a reasonably harmonious coexistence until the dawn of the second millennium. But that equanimity would not last forever. In 1001, the Afghani sultan Mahmud of Ghazna launched his first attack on the subcontinent, with the dual aim of destroying the infidels and looting their palaces and temples to fund his growing empire. The 1001 raid would be the first of sixteen distinct attacks over the next thirty years. Three years later, Mahmud had crossed the Indus; in 1008 he stormed the citadel of Kangra and walked away with 180 kilos of gold ingots and two tons of silver bullion.

Mahmud's greed was matched by his remorseless assault on the icons of the Hindu faith. (The word "iconoclast," now used as a largely approbative term to describe eccentrics, originally referred to destroyers of religious symbols.) Mahmud's armies had ultimately made it as far south as the Ganges plain by the time of his death in 1030. Within two centuries, Muhammad Ghuri would establish the Delhi sultanate in which, for the first time, the bulk of the Indian subcontinent was under Islamic control, where it would remain for five centuries.

The nature of the Muslim reign over India remains a highly contested question, even today. Some consider it the most devastating genocide in world history. The historian Fernand Braudel describes it in his *A History of Civilizations*:

> The Muslims could not rule the country except by systematic terror. Cruelty was the norm—burnings, summary executions, crucifixions or impalements, inventive tortures. Hindu temples were destroyed to make way for mosques. On occasion there were forced conversions. If ever there were an

uprising, it was instantly and savagely repressed: houses were burned, the countryside was laid waste, men were slaughtered and women were taken as slaves.

Other accounts paint a picture of a more tolerant Muslim rule, most notably under the Grand Mughals who came to power with the rise of Babur in 1526. At the apex of the Mughal dynasty—conventionally associated with the rule of Akbar the Great in the second half of the 1500s—India enjoyed a dynamic economy and limited religious discrimination. Akbar was himself a scholar of world literature; he appointed many non-Muslims to civil posts and eliminated a tax that specifically targeted Hindus. He even attempted to form a hybrid religion, known as the Din-i Ilahi, or "Divine Religion," that would incorporate elements from both Islam and Hinduism, though it never took hold.

The last Muslim leader to rule India without significant contest would come to power in 1658, within a few years of Henry Every's birth. His full imperial title was Abu Muzaffar Muhiuddin Muhammad Aurangzeb Alamgir. But to most of the world, he was known by a single name: Aurangzeb.

Imagine a split-screen vision of the late 1650s: an infant is born to an undistinguished family in the West Country of England, while five thousand miles away, the new heir to a dynasty ascends to the Peacock Throne for the first time. It is hard to imagine two lives with less in common, separated as they were by geography, culture, class, religion, and language. But as improbable as it might have seemed at the time, a series of events would eventually draw mighty Aurangzeb and Henry Every into violent conflict with each other.

That unlikely intersection had consequences that extended far beyond the scale of individual lives. A spectator in the late 1650s watching that split-screen view of Every's birth and Aurangzeb's ascension would have found it almost impossible to believe that the Islamic era in India was about to collapse, giving way to British imperial forces that would control the subcontinent for two more centuries. The British occupation of India is such a defining fact of the modern age that it is hard to imagine an alternate timeline. But if the story of Henry Every's life had played out differently, that occupation might not have happened at all.

4

HOSTIS HUMANI GENERIS

Algiers
Circa 1675

While Henry Every would eventually become the most notorious pirate in the world, he may well have begun his Royal Navy career attempting to rid the seas of the terrible scourge of piracy. According to a biography of Every by Adrian Van Broeck, young Every "set sail from Plymouth" aboard a "Fleet of Men of War that was then going to suppress the Nest of Pirates at Algiers." Following a narrative arc that would become increasingly common in the sea novels that flourished in the nineteenth century, Van Broeck has Every quickly making a name for himself on board. "Young Every shews an uncommon readiness in the practice of maritime affairs," he writes, "and not only gets into the esteem of the officers of his Majesty's ship the Revolution, which he served aboard, but of the Commadore Rear Admiral Lawson . . . having exerted an extraordinary Vigor and Sprightliness while Algiers was reduced to reason by the terror of the English Navy."

Elements of Van Broeck's account do have a basis in historical fact. A Vice Admiral John Lawson did in fact command a fifty-gun frigate called the HMS *Resolution*, and spent several years attempting to protect British merchant ships from the Barbary pirates that operated out of Algiers, Tunis, and Tripoli. The problem is that Lawson's main tour of duty in the Mediterranean took place in the early 1660s, and Lawson himself died in a naval battle with the Dutch off the coast of Suffolk in 1665; the *Resolution* was sunk in the St. James' Day Battle with the Dutch the following year. If Henry Every was in fact born in Newton Ferrers in 1659, he would have to have been an usually precocious sailor to have served with John Lawson in Algiers in the early 1660s. (Perhaps his "Vigor and Sprightliness" derived from the fact that he was only three years old at the time.) Of course, teenage boys were a regular presence on Royal Navy ships, and in Van Broeck's account, Every was born in 1653, which leaves open the slender possibility that he set sail with Lawson on the *Resolution* as a cabin boy in his eighth or ninth year. But that would have been uncommonly young even by the standards of the Royal Navy, and it seems unlikely that a child of that age would have made any impression whatsoever on an admiral, however "vigorous" he might have been.

A second HMS *Resolution*—this one a seventy-gun, third-rate ship of the line—launched in 1667 and was also deployed against the Barbary pirates in the late 1660s, though not with Lawson on board. But if we are to believe that Every was born in 1659, and ultimately participated in some kind of nautical battle in which "Algiers was reduced to reason by the terror of the English Navy," then the most likely scenario is that Every joined the navy in the early 1670s and participated in a series of attacks against the cities of the Barbary Coast during that period.

Whatever the actual chronology might have been, it does seem fitting that Every would have been inspired to join the navy by the promise of inflicting terror on the Barbary pirates. Growing up on the southern coast of England, the legendary corsairs of North Africa would have played a prominent role in the nightmares and folktales of Every's childhood. The Barbary pirates had been attacking British merchant ships in the Mediterranean for more than a century, but they also posed a far more immediate threat to the coastal communities of England and Ireland. In 1631, a Barbary pirate raid on the small Irish village of Baltimore in County Cork in the dead of night absconded with almost a hundred people, half of them children, all of whom were sold into slavery back in Algiers. Fourteen years later, two hundred forty English citizens living on the Cornish coast were captured and enslaved. (Many were ultimately ransomed by Parliament and returned to England years later.) Rumors held that as many as sixty Barbary men-of-war were actively prowling the English Channel, waiting for the opportunity to capture more product for the slave markets of Algiers and Tripoli. For most of the seventeenth century, an English or Irish family living near the coast confronted the real possibility that they might be hauled off without warning to a North African prison. A Committee for Algiers established by Parliament in 1640 estimated that as many as five thousand English citizens were enslaved in North Africa. Those numbers suggest that the odds of sudden enslavement by Barbary pirates were far higher for the average Devonshire resident than the odds of experiencing a terrorist attack in a modern-day Western city.

From the British perspective, these predations meant that the Barbary pirates were classified according to a venerable legal tradition, one of the earliest terms of international law: *Hostis humani generis*, Latin for "enemies of all mankind." Raiding coastal villages to kidnap families and sell them into slavery constituted a transgression that went beyond the usual offenses of criminal behavior. The

Barbary pirates had committed crimes against humanity itself, and thus warranted more extreme forms of punishment for their actions. For centuries, the classification of *Hostis humani generis* was reserved exclusively for pirates—Every and his men would find themselves condemned with that phrase two decades after the British Navy "reduced Algiers to reason"—in part because the pirates committed acts of atrocity that went beyond the usual boundaries of criminal behavior, but also because they committed most of those acts in international waters, where legal jurisdictions were by definition blurry. Declaring that pirates were "enemies of all mankind" gave local authorities on land the legal justification to try them for their crimes, even if those crimes had taken place on the other side of the world. But in the twentieth century, *Hostis humani generis* would be extended to a broader group of outlaws: war criminals, torturers, and terrorists all found themselves under its ancient umbrella. In the immediate aftermath of 9/11, the Justice Department lawyer John Yoo invoked the tradition of *Hostis humani generis* to justify the extreme treatment of enemy combatants as part of the war on terror. The legal groundwork for the abuses of Guantanamo Bay and Abu Ghraib was first laid down to address the unique transgressions of pirates on the open sea.

There was no shortage of hypocrisy in seventeenth-century Britain condemning the Barbary pirates as enemies of all mankind. Some of the most notorious pirates in the world had been English, and had gone about their business with the full endorsement of the Crown. English law during this period attempted to erase this apparent contradiction through a technical loophole, based on the distinction between pirates and privateers. In their actions, privateers seemed almost indistinguishable from pirates: they sacked towns, captured treasure, and seized ships, torturing and killing along the way. But they did so with the blessing of their government, usually in the form of a "letter of marque" that gave them the authority to attack vessels belonging to other nations. "In return for this legal protection," the historian Angus Konstam writes, "the state that had issued the letter of marque usually received a percentage of the profits. As long as they abided by the rules and attacked only the enemies of the state listed on their letter of marque, privateers could not be hanged as pirates, condemned to a lifetime of servitude in the galleys, or simply killed outright." Conventionally, the privateers were only allowed to attack ships belonging to nations that were officially enemy states, under a formal declaration of war. But the lines were often blurred, and the privateers, who had developed a taste for the buccaneer's lifestyle, were ill-inclined to give it up when the official hostilities ended. "Privateers in time of War," the early pirate historian Charles Johnson observed in his *General History of the Pyrates*, "are a Nursery for Pyrates against a Peace."

Privateering as a formal assignation dates back to the reign of Edward I. British merchant ships that had been attacked by pirates were granted "Commissions of Reprisal"—the forebear of the letter of marque—which gave them the right to capture non-British merchant ships themselves. Technically, the arrangement was designed to be a strict tit-for-tat: the privateers were supposed to only seize ships flying the colors of the pirates that had originally stolen from them. But in practice, the privateers were less discriminating, and often hauled in far more treasure than they had originally lost.

Privateering came into its own in the 1500s as England's relationship with Spain grew increasingly hostile, a period in which "legitimate trade, aggressive mercantilism, and outright piracy commingled and coalesced," as the historian Douglas Burgess describes it. With Spanish galleons transporting untold riches of silver, gold, and spice from the Americas back to Seville, and the stigma of piracy removed by the letter of marque, privateering became a career path for a more respectable class of men, most famously Francis Drake, the son of a Devonshire minister, who circumnavigated the globe in the late 1570s and led a series of devastating attacks on Central American ports, accumulating enough wealth and prestige through his adventures to be knighted by Elizabeth I and acquire the stately manor house Buckland Abbey in Devon, now preserved by the National Trust. As Burgess writes, "Drake's colossal success not only made him a hero, it made him a prototype—the standard by which all future pirates would be judged and by which they judged themselves."

All of this history meant young Henry Every would have had two distinct models of piracy in his mind as he left Plymouth with the Royal Navy: the murderous Barbary pirates, living outside the boundaries of human decency, enemies of all mankind; and the dashing figures of Drake and other successful privateers—esteemed men who had lived lives of great adventure and risk, and who had profited mightily from their labors. To be a pirate meant that you were simultaneously beneath contempt and on a thrilling road to respectability—even to knighthood. Those two polarities were maintained for at least a century without much cognitive dissonance for an obvious reason: the Barbary pirates were (mostly) North Africans, attacking innocent British families, while Drake and his peers were sacking Spanish settlements in the New World. That the former should seem monstrous and beyond the pale and the latter worthy of knighthood was simply a case of rooting for the home team.

Henry Every would have had no way of knowing it, in those first years of his naval career, but his actions would ultimately compel those two models of piracy to collide with each other, forcing the British to confront the possibility that one of their celebrated buccaneers might just be a monster after all.

5

TWO KINDS OF
TREASURE

Surat, India
August 24, 1608

The merchant galleon *Hector* that dropped anchor at the mouth of the Tapti River on the western coast of India in late August of 1608 had been at sea for more than a year, sailing from London around the Horn of Africa, with extended stops in Sierra Leone and Madagascar to reprovision. The sight of a European trading vessel on the Tapti would have been nothing new to the Indians living along the river; just fourteen miles upstream lay the harbor town of Surat, the epicenter of Red Sea trade. But a discerning observer would have noticed something unusual about the *Hector*. In an age where the Portuguese held a monopoly over European commerce with India—a tradition that had begun with Vasco da Gama's famous voyage of 1499—the arrival of the *Hector* marked a critical turning point in the relationship between India and Europe. It was the first ship flying British colors to reach the shores of the Indian subcontinent.

On board the *Hector* was a representative of the East India Company named William Hawkins, who had been dispatched by the company to investigate the possibility of opening new avenues of trade with India. The general lessening of tensions after the Treaty of London in 1604, which ended the Anglo-Spanish War, had led the company to believe the Portuguese might tolerate other traders in the Indian ports they controlled, and the company's recent troubles in the Spice Islands made its directors particularly eager to secure new markets. Hawkins carried with him a letter from King James to the Grand Mughal Jahangir, requesting that the sultan grant "such liberties of traffique and privileges as shall be resonable both for their securitie and proffit."

In Surat, Hawkins was initially informed that the local governor was "not well" and would not be able to meet with him. (In his diary, Hawkins notes that he suspected the cause was an opium stupor, not ill health.) Instead he was welcomed by the *shahbander*, or port master, of Surat. "I told him that my comming was to establish and settle a factory in Surat," Hawkins recorded in his diary, "and that I had a letter for his king from His Majesty of England tending to the same purpose, who is desirous to have league and amitie with his king, in that kind that his subjects might freely goe and come, sell and buy, as the custome of all nations is; and that my ship was laden with the commodities of our land which, by intelligence of former travellers, were vendible for these parts."

At first, Hawkins's overtures seemed to induce a favorable response. The morning after his meeting with the port master, he learned that the governor was now well enough to receive him. Dressed in an elaborate scarlet taffeta outfit embroidered with silver lace, designed back in London specifically to create an aura of ambassadorial distinction, Hawkins presented the governor with gifts and repeated his desire to establish a commercial relationship with

Jahangir's regime. "With great gravity and outward shew of kindness he entertained me," Hawkins wrote, "bidding me most heartily welcome, and that the countrey was at my command." The welcome, however, would prove to be short-lived. A customs officer named Muqarrab Khan seized some of the "vendible commodities" that Hawkins had hoped to sell to the merchant community in Surat; the rest fell into the hands of the Portuguese, who also captured most of Hawkins's crew, declaring that "Indian seas belonged exclusively to Portugal." Dodging several murder plots against him, Hawkins escaped with two men and began the long overland trek to the capital city of Agra, in the hopes that the Grand Mughal himself might be more open to his proposition of "amitie" with King James and the traders of the East India Company.

Hawkins's persistence paid off in the end. In Agra, he found a city of staggering architectural grandeur and opulence, with forts and palaces constructed out of the region's distinctive red sandstone. (The ivory white marble domes of Agra's most famous structure, the Taj Mahal, would not be built for another three decades.) Lush tropical gardens, replete with octagonal pools, pavilions, and mausoleums, lined the Yamuna River. At the end of a journey that entailed "much labour, toyle, and many dangers," Agra must have seemed like something out of a dream.

Hawkins's reception at the court of Jahangir proved to be far more successful than his initial encounters in Surat. Having lost almost all of his "vendibles" to Surat's port master and the Portuguese, Hawkins had only "a slight present" of cloth to offer as tribute to the Grand Mughal. But the missive from King James resonated with Jahangir: "He spake unto me in the kindest manner," Hawkins later wrote, "granting and promising me by God that all what the King had there written he would grant and allow with all his heart, and more if His Majestie would require it." The two men then discovered that they shared a common language, Turkish, and in a long conversation about the different nations of Europe, they began a complicated friendship that would last for almost four years.

Hawkins's travels from Surat had stripped him of almost all his belongings, and nearly cost him his life multiple times, but overnight, under the graces of the Grand Mughal, he was immediately elevated to a far more lavish lifestyle. Jahangir declared that Hawkins should become a "resident Ambassador" at Agra. "He was made captain of four hundred horse, with a handsome allowance, was married to an Armenian maiden, and took his place among the grandees of the court," according to historian William Foster. Discarding his ragged taffeta uniform, Hawkins began to dress in the "garb of a Muhammadan noble."

During his stay at Agra, Hawkins made an important contribution to a venerable genre of "Orientalist" literature: the European marveling at the astonishing opulence of the Indian elite. The entire second half of Hawkins's diary of his years in India is devoted to an exacting inventory of the Grand Mughal's extravagant lifestyle. "His treasure is as followeth," Hawkins announces, and then proceeds to list his "coins of gold," his "jewells of all sorts," his "jewels wrought in gold," "all sorts of beasts," all the way down to the jewel-encrusted furnishings of the royal court:

Of chaires of estate there bee five, to say, three of silver and two of gold; and of other sorts of chaires there bee an hundred of silver and gold; in all an hundred and five. Of rich glasses there bee two hundred. Of vases for wine very faire and rich, set with jewels, there are an hundred. Of drinking cuppes five hundred, but fiftie very rich, that is to say, made of one piece of ballace ruby, and also of emerods [emeralds], of eshim, of Turkish stone [turquoises], and of other sorts of stones. Of chaines of pearl, and chaines of all sorts of precious stones, and ringes with jewels of rich diamants, ballace rubies, rubies, and old emerods, there is an infinite number, which only the keeper thereof knoweth.

Hawkins's amazement at the sheer scale of the Grand Mughal's treasure is a reminder of an important conceptual frame that shaped encounters between Europe and India during this period: many Europeans assumed that India, of the two cultures, was the wealthier one. Measured purely by the standard of the production of luxury goods, it was no contest. Economists now believe that the per capita GDP in India was likely close to that of Europe circa 1600, but the concentration of wealth among the ruling elite was far greater in the subcontinent. At the high end of society—in the palaces, the royal gardens, all the outward appearances of wealth and civilization—India appeared to most European visitors to be the more advanced culture.

In his description of Jahangir's own personal dress, Hawkins hints at the origin of the Grand Mughal's vast treasure:

He is exceeding rich in diamants and all other precious stones, and usually weareth every day a faire diamant of great price. . . . He also weareth a chaine of pearle, very faire and great, and another chaine of emeralds and ballace rubies. Hee hath another Jewell that commeth round about his turbant, full of faire diamants and rubies. It is not much to bee wondered that he is so rich in jewels and in gold and silver, when he hath heaped together the treasure and jewels of so many kings as his forefathers have conquered, who likewise were a longtime in gathering them together, and all came to his hands. Againe, all the money and jewels which his nobles heape together, when they die come all unto him, who giveth what he listeth to the noblemens wives and children; and this is done to all them that receive pay or living from the King, India is rich in silver, for all nations bring coyne and carry away commodities for the same; and this coyne is buried in India and goeth not out.

This coyne is buried in India and goeth not out. The line could serve as a slogan for India's economic program. From the Roman days on, India displayed little interest in the products that Europeans offered to trade in return for the spices and fabrics and other goods that were so highly valued by European consumers. If the citizens of Europe wanted pepper on their table or calico on their skin, they had to pay for it in bullion. But instead of turning that wealth into working capital, most of it went into the grand displays that had so dazzled Hawkins and his contemporaries. "India had long been 'an abyss for gold and silver,'" the historian John Keay writes, "drawing to itself the world's bullion and then nullifying its economic potential by melting and spinning the precious metals into bracelets, brocades and other ostentatious heirlooms." The gold arrived in India as a form of currency, but it was ultimately turned into an ornament, the equivalent of winning the lottery and decorating your house with wallpaper made of hundred dollar bills. Yet in the early days of the seventeenth century, the economic model that the Mughals had pursued seemed undeniably to be working. If your goal was to assemble a vast fortune, the path pursued by the Grand Mughals—and, it should be said, by King James and the other European monarchs— appeared to be the most viable option.

But William Hawkins was more than just a representative of King James. He was also a kind of emissary from the future. He was there as a representative of both a nation-state *and* a private corporation, the East India Company.

This is what ultimately makes that meeting between Hawkins and Jahangir so important, from the long view: it was the first point of contact between two different strategies for accumulating treasure. The first was an old trick, almost as old as agriculture: declare yourself emperor/king/mughal and extract rents from everyone beneath you in the form of taxes and tariffs. This approach had a long track record of success. Jahangir's "infinite number" of jewels and precious stones marked the outer boundaries of that strategy's returns, but they were not unusual at that moment in history. If you wanted to join the super-rich in 1600, the entirely fictitious notion of "royal blood" was the main point of entry. But that was about to change. Within a few centuries, the monarchies would become high-class pensioners, living off a lavish but ever-declining dole. The real money would be made elsewhere.

Mostly it would be made by joint-stock corporations and their shareholders. Royal families barely make an appearance on the Forbes 100 these days; today the upper echelons of the super-wealthy are composed almost entirely of people who have participated in public or semipublic offerings of stock in corporations, either as founders (think Bill Gates and Jeff Bezos) or as investors (think Warren Buffett). As a representative of the East India Company *and* an emissary for King James, Hawkins was in the service of two masters. He swore loyalty to a British king who, in terms of the feudal economic model that sustained him, bore a meaningful resemblance to Jahangir. But Hawkins was also a representative of the East India Company, by most accounts the first joint-stock corporation in the history of human commerce.

Queen Elizabeth I had granted a charter to form "one body corporate" with the formal name "The Governor and Company of Merchants of London trading into the East Indies" on December 31, 1600. Publicly traded shares, giving investors a stake in a company's overseas ventures, were offered to a wide swath of wealthy British society: "earls and dukes, privy councilors, judges and knights, countesses and ladies of rank, widows and maiden ladies, clergyman, merchants, tradesmen and merchant strangers." Before the East India Company, if you wanted to capture some of the value of these emerging global trade networks, you had to set sail with Francis Drake or one of his contemporaries. (Or be a member of the royal family.) Joint-stock offerings gave you a piece of the action without forcing you to leave your London coffeehouse. All you had to do was buy a few shares.

Initially, the company issued what was called "terminable stock," based on single voyages or sometimes clusters of three or four voyages. The company would raise funds for a voyage, say, to the Spice Islands, and if the voyage proved to be successful, the profits were distributed among the shareholders based on the size of their original investments. But by the middle of the 1600s, the company had shifted to the model that most corporations employ today: issuing permanent stock that reflected an investment in all the company's current and future ventures. This innovation had two crucial benefits, and one interesting side effect. Raising funds from a large pool of investors meant that, for the first time, business ventures with massive fixed capital costs—for instance, building and transporting ships across the world to purchase goods to sell back to British consumers— could raise enough money from private citizens that they could operate without the direct oversight and backing of the state. (At the height of its power in India during the 1700s, the East India Company would effectively function as its own state apparatus, with standing armies and corporate officers controlling vast stretches of the subcontinent.) Secondly, by distributing the investment across many individuals, the company minimized the risk of any individual venture. If a ship sank on its way back from India, the loss would be felt widely among the investor classes back in London. But because the voyage had been funded by many small contributions—instead of by, say, one royal patron—the impact of the sinking was less catastrophic.

The side effect was that issuing publicly traded stock created a secondary market for the shares themselves. Those shares would rise and fall in value as the fates of the East India Company rose and fell over the 1600s. But mostly they rose. Between 1660 and 1680, shares in the company quadrupled in value, driven in large part by the craze for calico and chintz that swept through the London elite during that period. (By the 1680s, the East India Company was importing almost two million pieces of cloth annually, dwarfing the trade in spices that had originally inspired Elizabeth to grant the corporate charter.) The increase in share value constituted a genuinely new kind of wealth. The corporation itself made money in the time-honored tradition of merchants, dating back to the Islamic traders and beyond: they bought low and sold high, and their profits reflected the delta between those two prices. Some of those profits flowed back to the investors in the form of dividends. But the trading of shares created a second form of wealth, one that, in the long run, proved to be more lucrative. You made money investing in the East India Company not just because the company turned a profit, but because other investors thought your shares were worth more than you had paid for them.

And so that meeting between the two men in Agra, in the spring of 1609, marks an early milestone in the transition from one regime of wealth accumulation to another, a transition that would spread from London and ultimately sweep across the entire planet, as the joint-stock corporation became, by the twentieth century, the dominant form of organizing economic activity, in the private sector, at least. Hawkins wouldn't have looked all that impressive in his tattered taffeta next to Jahangir's "chaines of emeralds and rubies." But he had the future on his side.

Despite Jahangir's apparent fondness for Hawkins, the Portuguese managed to intervene to keep the English out of the Indian trade for several years. Hawkins left Agra in 1611, and died at sea shortly thereafter. It was not until 1612 that the Grand Mughal granted the East India Company a permit to open a factory at Surat. Hawkins's successor, Thomas Roe, secured a missive from Jahangir to King James that made the terms explicit:

> I have given my general command to all the kingdoms and ports of my dominions to receive all the merchants of the English nation as the subjects of my friend; that in what place soever they choose to live, they may have free liberty without any restraint; and at what port soever they shall arrive, that neither Portugal nor any other shall dare to molest their quiet; and in what city soever they shall have residence, I have commanded all my governors and captains to give them freedom answerable to their own desires; to sell, buy, and to transport into their country at their pleasure.

The factory at Surat marked the first foothold on Indian soil for this fledgling new corporation. From that small port city, where the victims of Henry Every's predations would seek their revenge almost a century later, the British would steadily extend their domain, most famously in their settlements at Bombay and Madras. Before long, the entire subcontinent would be under the rule of the East India Company.

6

SPANISH EXPEDITION SHIPPING

East London
August 1693

I f Henry Every spent his childhood living in fear of being abducted into slavery by Barbary pirates, the experience seems to have not had much of an influence on his subsequent ethical feelings about the institution of slavery itself. The first clear reference to Every in the historical record—after his Royal Navy engagement as a young man—comes from an agent of the Royal Africa Company (RAC), Thomas Phillips, who reported in 1693 that Every had taken up a career as a slave trader, working for the governor of Bermuda. At the time, the RAC claimed a monopoly on all English slave trade in the region. British history often conveniently neglects the sheer scale of the company's involvement in the slave trade during this period, focusing instead on the fact that slavery was largely abolished in England itself—if not in her colonies—by the late 1700s. But as the historian David Olusoga observes, "the Royal African Company transported more Africans into slavery than any other British company in the whole history of the Atlantic slave trade . . . around a hundred and fifty thousand men, women and children passed through the company's coastal fortresses on their way to lives of miserable slavery." According to the RAC agent Phillips, Henry Every had built a career for himself in the early 1690s as an interloper, working outside the official monopoly of the RAC, sometimes capturing the English traders themselves along with their African captives. Typically for Every, even Phillips's unmistakable reference contained two aliases for the man. "I have nowhere met the negroes so shy as here," the agent wrote, "which makes me fancy they have had tricks played on them by such blades as Long Ben, alias Every, who have seized them and carried them away."

But Every would not truly enter the main stage of history until the following year. An affluent investor and MP named James Houblon had gathered a group of relatives and tradesmen in London to fund a new speculative venture. One of twelve children, Houblon was a member of a prestigious London family with extensive ties to the East India Company; his brother John would be the first governor of the Bank of England. (His portrait appeared on a £50 note issued during the 1990s.) The enterprise—which went under the name Spanish Expedition Shipping—planned to assemble a squadron of ships loaded with guns and cannon, which would then set sail for the West Indies, trading some of the arms to the Spaniards there. Houblon had made a small fortune importing Spanish wines and other foodstuffs, and he had used his connections with Madrid to secure a trading and salvage license from Carlos II. The Spanish Expedition would make its real money, Houblon and his investors believed, by salvaging treasure from sunken Spanish galleons in the Caribbean. Led by an Irish admiral named Don Arturo O'Byrne, the fleet consisted of four vessels: the *James*, the *Dove*, the *Seventh Son*, and the flagship, a sleek newly constructed forty-six-gun "ship of force" named the *Charles II.*

Houblon had commissioned the *Charles II* specifically for the Spanish Expedition, the ship constructed in the East London dockyards near where the East India Company built its own vessels. In addition to the extensive armaments onboard the ship, the *Charles II* was uncommonly fast and agile. Documents that survive from the period suggest that Houblon was particularly enamored of the ship. He called her a "great merchant-man . . . a stout frigate of forty guns and an extraordinary sailer." Houblon and the other investors had gone to the expense of building such an intimidating flagship so that the squadron could defend itself against any would-be attackers off the coast of Spain or in the West Indies. Just a year earlier, John Houblon had written an impassioned note to the Board of Trade, begging for a convoy of men-of-war to accompany his merchant ships returning from Lisbon. More than a dozen Barbary pirate ships were prowling the coast, he warned, and "French privateers off of the coast of Portugal intercept[ed] and [took] several English and Irish ships." Several months later, James Houblon sounded a similar note in a petition to the Privy Council, asking for naval support for ships he had dispatched to trade with Spain. "The ships will be very richly loaden with Spanish wooll and [considerable monies], and other rich comodityes," he wrote. "Wherefore they humbly pray you Lords be pleased to order a speedy convoy to fetch home the said ships, suitable to the richness of the fleete and the Danger they will run." By investing up front in their "stout frigate of forty guns," Houblon and the other Spanish Expedition investors would no longer need to plead for help from the Privy Council. The *Charles II* could outrun whatever danger it might encounter on its journey.

Houblon and his fellow investors recruited their crew by promising regular wages, with a month's advance paid up front on signing up for the expedition—a far more generous deal than anything the Royal Navy would have offered. Because the Spanish Expedition would quickly prove to be an utter failure—at least in terms of its initial objectives—the venture ultimately triggered a number of lawsuits that give us a documentary record of the financial package granted to some of the crew. One high-ranking sailor on the *Dove* was offered four pounds, ten shillings per month, with a total package of £82—roughly the equivalent of $20,000 in today's currency—for the entire voyage. With such well-heeled backers, the food and grog aboard the ship promised to far exceed what you would have found in the mess of a Royal Navy ship (at least for the ordinary seamen on board). Prospective crew would no doubt have expected to supplement their wages with some of the eventual plunder, making the whole proposition even more enticing.

In August 1693, Houblon paid a personal visit to the squadron as it took on provisions while anchored on the Thames. Houblon promised that the families of the crew would be compensated during their long journey, and wished them bon voyage. Shortly thereafter, the four ships raised anchor and began sailing toward the mouth of the Thames and out onto the open sea.

Roughly two hundred men were aboard the four vessels. The fleet was notable for the experience of its officer class. John Knight, captain of the *Dove*, was reputed to be a "sober, diligent, and knowing man" who had already commanded a number of ships in voyages to the West Indies. The pilot aboard the *Charles II* was a veteran Spanish navigator named Andres Garsia Cassada. The original captain of the *Charles II* was John Strong, who had led a successful salvage operation off the coast of modern-day Haiti several years before. But Strong would die early in the voyage, replaced by an alcoholic mariner by the name of Charles Gibson.

Below the officer class, the biographical details get blurrier. There was Thomas Druit, first mate of the *James*; Joseph Gravet and David Creagh, second mates on the *Charles*, along with Henry Adams, who would ultimately be promoted to quartermaster on the ship. The steward on board the *James* was a fifty-year-old sailor named William May, a "very sickly man," by his own account, who had served his "King and Country for thirty years." Also aboard was forty-nine-year-old coxswain John Dann, originally from Rochester, along with a forty-five-year-old sailor from Newcastle upon Tyne named Edward Forsyth. At the other end of the generational spectrum were an ambitious but unseasoned young sailor in his early teens named Philip Middleton, a seventeen-year-old Londoner named John Sparkes, and William Bishop, an eighteen-year-old sailor on his first voyage at sea, who later claimed to have been forced into service on the *James* against his will.

One of the most intriguing characters to sign up for the Spanish Expedition was the second mate of the *Dove*, a veteran seaman and scientist named William Dampier. In his early forties, Dampier had already circumnavigated the globe once, in a rambling series of voyages that lasted almost the entire decade of the 1680s. (He would go on to become the first man in history to circumnavigate the earth three times.) A few years after the Spanish Expedition's spectacular demise, Dampier published a memoir of his travels called *A New Voyage Round the World*. While the book was surprisingly silent about Dampier's connection with the *Charles II*—at the time, the subject of much speculation in the popular press—the book went on to be a bestseller, and helped inaugurate a tradition of travel writing that would become one of the eighteenth century's most popular nonfiction genres. Novelists, as well, were deeply influenced by Dampier's tales: both Daniel Defoe's *Robinson Crusoe* and Jonathan Swift's *Gulliver's Travels* drew heavily from *A New Voyage Round the World*.

The travelogues so impressed the admiralty that Dampier was ultimately granted command of a warship, HMS *Roebuck*, on which he made a historic voyage to Australia, where he documented the continent's unique flora and fauna. His botanical studies—as well as pioneering work that he produced on the connection between trade winds, tides, and ocean currents—made him a role model for Charles Darwin, who read extensively from Dampier's travel narratives and naturalist studies during the voyage of the *Beagle*. Today, a portrait of William Dampier hangs in the National Portrait Gallery in London.

The fact that Dampier was so circumspect about his time aboard the *Charles II*—in books that otherwise documented his travels in exhaustive detail—was likely a strategic move. Throughout his career, Dampier operated at the blurred boundaries between pirate and privateer, maintaining a level of legitimacy that would keep him in good graces with the Royal Society and the British Admiralty. An association with the *Charles II* threatened to undermine that delicate positioning, thanks to another crew member who had joined the Spanish Expedition, one whose fame would for a time far eclipse that of William Dampier: the first mate of the *Charles II*, Henry Every.

Every was likely in his late thirties, tall and in impressive physical condition. He had striking gray eyes and wore a "light coloured Wigg," according to one of his shipmates. He was embarking on what must have seemed at the time to be a promising but not particularly exceptional voyage. He would have had at least a dozen comparable expeditions under his belt at that point in his life. Did First Mate Every have an inkling, standing on the deck as the *Charles II* made its way down the Thames, that this mission might mark the turning point in his life? On that question, as on so many questions about young Henry Every's life, the historical record is silent. But one fact is clear: Every would end the journey with an elevated rank, from first mate to captain, from an anonymous sailor to the world's most notorious criminal.

For five other members of Spanish Expedition Shipping, the voyage would end with a different kind of elevation: hanging from a noose at Execution Dock.

7

THE UNIVERSE CONQUEROR

Delhi

September 1657

nce a year, the graves of the royal couple entombed in the Taj Mahal—the Grand Mughal Shah Jahan and his queen, Mumtaz Mahal—are opened to the public for three days to commemorate the death of Shah Jahan, who built the epic mausoleum as a tribute to his late wife. For those few days, known as the *urs*, the usual entry fees charged to visit the Taj Mahal are waived and a vast throng of visitors descends on the site. The Khuddam-e-Roza (the shrine's caretakers) carry an immense, multicolored cloth known as a *chadar*—which can be over 800 meters long—to mark the beginning of the *urs*. The colors of the *chadar* represent the many religions of India, and the whole ceremony has a sublime elegance that befits a shrine built as a "monument to love."

While the death of Mumtaz Mahal inspired one of the great wonders of the world, the death of Shah Jahan himself had a less uplifting arc. The Grand Mughals of India shared many structural traits with

European monarchies—lifetime autocratic rule justified by divine right, a lavish lifestyle supported by taxes and tariffs, surrounded by the proto-bureaucracy of court society—but the Muslim rulers did not share the primogeniture rights of succession that had become nearly ubiquitous across Europe in the feudal age. When a Mughal died, his power did not pass directly to his eldest son. Each male offspring was considered to have a legitimate claim to the throne. The ambiguity inherent in these succession rights meant that the death of a Mughal was often immediately followed by an outbreak of royal fratricide, as the surviving sons battled one another to inherit their father's title. For all his subsequent expressions of romantic love, Shah Jahan was not above a ruthless play for power in his younger years. When his father, Jahangir, died in 1627, almost two decades after his fateful meeting with William Hawkins, Shah Jahan executed his brother and two nephews in his bid to become Grand Mughal.

Fifty years later, Shah Jahan's reign would come to an even more brutal end.

Sometime in September of 1657, the sixty-one-year-old Mughal began suffering from a debilitating illness while at court in the new capital city of Delhi. The famous seventeenth-century historian Khafi Khan—who would later write the definitive Indian account of Henry Every's crimes—claimed in one of his chronicles that Shah Jahan had developed a condition known as a "strangury," which involves painful, frequent urination, though other accounts report that he suffered from acute constipation. Before long, the Mughal had developed a high fever, and rumors of his imminent demise spread across the land, producing, according to Khan, "much derangement in the government of the country." This was almost certainly an understatement.

With their father near death, the sons of Shah Jahan immediately began preparing to stake their claims to the throne. Two of the brothers, Prince Shuja and Prince Murad, commenced printing coins with their own likenesses; Shuja even staged a coronation, blithely ignoring the fact that his father was still alive. Both brothers must have known they had slim chances of actually succeeding Shah Jahan, given his fondness for his eldest son, Prince Dara, who was based in Delhi and had served as a kind of surrogate spokesman for his father for years. (Dara also enjoyed the support of his sister Jahanara, who had her own sphere of influence within the Mughal court.) But Dara had one critical vulnerability: his Islamic bona fides were suspect. In the tradition of his distant ancestor Babur, Dara had extensive social and intellectual connections with Sufis, Hindus, and Christians alike. He had even publicly argued that "the essential nature of Hinduism is identical with that of Islam." Among orthodox Muslims and the religious scholars known as the ulema, Dara was seen as almost a heretic. Fearing a retreat back to the more tolerant era of Babur, they threw their weight behind Dara's far more pious brother, Aurangzeb.

In late spring of 1658, the struggle to succeed Shah Jahan culminated in a fierce battle between Dara and Aurangzeb on a wide desert plain eight miles outside Agra. The two brothers commanded cavalries riding as many as fifty thousand horses, supported by a battery of war elephants. Along with rifles and cannons, the artillery men had portable rockets made with bamboo rods and iron points. Dara and Aurangzeb commanded their squadrons from their respective howdahs, ornate carriages strapped onto the backs of elephants. At one point in the melee, one of Dara's Rajputs (members of India's elite warrior caste) slashed his way through Aurangzeb's guard with only a sword and attempted to cut the cables that secured the howdah to the prince's elephant. According to Khafi Khan—whose vivid account of the battle stands as one of the great literary achievements in the history of military writing—Aurangzeb "became aware of this daring attempt, and in admiration of the man's bravery, desired his followers to take the rash and fearless fellow alive, but he was cut to pieces."

Across the battlefield, Dara was having a crisis of confidence. "Seeing so many of his noble and heroic followers killed and wounded, [he] was much affected," Khan wrote. "He became distracted and irresolute, and knew not what to do." It was the wrong time to waver in second thoughts. As Dara weighed his increasingly limited options from his howdah, a rocket launched by Aurangzeb's men made a direct hit on the outer edge of the carriage. Unhurt but dazed, Dara dismounted the elephant—"without even waiting to put on his slippers," Khan noted drily—and took to the saddle of a nearby horse. The sight of the commander's elephant swaying through the carnage, its howdah unoccupied, only demoralized the troops further. The last straw seems to have been a moment straight out of Peckinpah or Tarantino: as an attendant to Dara reaches up to the dumbfounded prince to hand him a quiver of arrows, an inbound cannonball rips between the two men, severing the attendant's hand at the wrist. That was it for Dara. "Beholding the dispersion of his followers, and the repulse of his army, prizing life more than the hope of a crown, [he] turned away and fled."

With Dara on the run, and his other rivals either dead or in hiding, Aurangzeb staged a quick coronation for himself in 1658, while still in pursuit of his older brother. Attempting to turn public opinion against the popular prince, Aurangzeb denounced Dara as an infidel, and encouraged locals to report any sightings. Reduced from an heir apparent living in unimagined opulence to a fugitive wandering the desert with his wife and daughter and only a few servants, Dara slunk westward from Agra to Gujarat. In the summer of 1659, Aurangzeb held a second coronation, this time on a much grander scale, with festivities lasting more than two months. Elaborate fireworks displays lit up the Delhi sky at night, and Aurangzeb distributed awards to thousands of subjects, no doubt trying to win over a court society that had long assumed Prince Dara would succeed Shah Jahan. He began referring to himself using the title Alamgir—"Universe Conqueror."

While Aurangzeb celebrated in Delhi, Dara made a last-ditch attempt to escape to Persia through the Bolan Pass. He took refuge in the foothills of the Central Brahui mountains, at the estate of Malik Jiwan, a local zamindar (well-to-do landowner). Shortly after their arrival, Dara's wife collapsed and died from dysentery. For Dara, the walls seemed to be closing in. "Mountain after mountain of trouble thus pressed up on the heart of Dara," Khan wrote. "Grief was added to grief, sorrow to sorrow, so that his mind no longer retained its equilibrium." Jiwan initially offered to help escort Dara across the Bolan Pass, but then seems to have had a change of heart. (Or Jiwan had been playing the long game from the beginning.) As Dara set off for his mountain exodus, a gang of "ruffians and robbers" descended on him, apparently at Jiwan's instruction. The zamindar sent word of the capture back to Aurangzeb, still immersed in the concerts and pyrotechnics of his second coronation. The newly ascendant Mughal had Dara transported back to Delhi, where the prince was placed on the back of a "dirty female elephant" and carried through the streets in chains. The spectacle appears to have backfired; rather than brand Dara as an apostate, the public shaming brought out a vast throng of supporters. The sight of ordinary Delhi citizens weeping for their fallen prince convinced Aurangzeb that "for public peace and reasons of the State, [it would be] unlawful to allow Dara to remain alive."

Dara was executed on August 30, 1659. His head was allegedly presented to his brother on a plate.

By the time Henry Every set sail on the Spanish Expedition,

Aurangzeb had occupied the Peacock Throne for more than three decades, a reign that shared many qualities with his violent ascent to power: aggressive military action coupled with a return to religious orthodoxy. Aurangzeb had abandoned the scholarly pluralism of Babur, along with Shah Jahan's commitment to pioneering architecture. (Aurangzeb's primary architectural legacy lies in the dozens of Hindu temples he destroyed.) Hindu pilgrims and all non-Muslim merchants had been subjected to new taxes. Extensive military conquests had extended his kingdom to include most of modern Pakistan, Afghanistan, and Bangladesh. As First Mate Every made his way down the Thames aboard the *Charles II*, more than 150 million people lived in territories controlled by the Grand Mughal. (At the time, the entire population of Europe was likely less than one hundred million.) Aurangzeb was almost certainly the wealthiest man in the world.

The Universe Conqueror's ascent to power was marred by one crucial asterisk. His father's terminal illness turned out not to be terminal at all. Shah Jahan lived for another eight years after his son clawed his way onto the Peacock Throne. That was eight years too many for Aurangzeb. He condemned his father to spend the rest of his life imprisoned in the Red Fort at Agra, with only a distant view of the Taj Mahal through his cell window to remind him of his former omnipotence.

8

HOLDING PATTERNS

A Coruña, Spain
Winter 1693-1694

If the crew aboard the *Charles II* left England in good spirits, anticipating a profitable adventure at sea, their enthusiasm would prove to be short-lived. The plan had been to make a quick, two-week voyage to the Spanish port city of A Coruña—known to the British as "The Groyne"—taking on supplies and securing additional paperwork there, before riding the trade winds to the West Indies. But for reasons that have never been entirely clear, the voyage to Spain ended up taking five months instead. This was only the first of several misfortunes to befall Spanish Expedition Shipping. Anchored in A Coruña, they were told that the paperwork required for the voyage was still en route from Madrid. As the weeks passed with no sign of the missing documents, the crew's unrest at the delay was aggravated by the fact that their promised semiannual salaries had yet to be paid. A petition delivered back to James Houblon was met with an order to lock the petitioners in the ship's brig.

Even in the best of times, life aboard a seventeenth-century privateering ship was a challenging and claustrophobic experience. The fact that a community of a hundred or more people could survive on the open seas for months at a time, in a vessel with dimensions not much larger than a tennis court, should go down as one of the great achievements in our long history of creating life-sustaining habitats in fundamentally inhospitable environments. A ship like the *Charles II* was as impressive, in its own time, as the International Space Station is in ours. Carrying forty-six guns on board, the *Charles II* was likely a little more than a hundred feet from bow to stern, with a galley beam measuring around thirty feet across its width. Counting the three main decks of the vessel, the captain's quarters at the stern, and the entire space below deck, the *Charles II* would have offered about six thousand square feet to house more than a hundred men, along with armaments, cargo, and enough food and drink to keep those hundred men alive for months at a time. For the captain and a few select officers, the onboard experience could be reasonably civilized; the stern windows in the captain's relatively spacious cabin looked out onto the open water, and his private galley enabled him to dine with his officers in style, even if the food was not up to the standards an affluent European might expect on land.

But belowdecks, it was another matter. In a ship with a hull loaded with cargo, the living quarters for the non-officer crew might be less than a thousand square feet. Ceiling heights were often less than five feet. Imagine squeezing yourself into a typical one-bedroom apartment with a hundred other men, with ceilings a foot shorter than the height of an average adult male, and somehow trying to get a good night's sleep. Only this apartment is prone to lurch violently from side to side in intense ocean storms without a single window to give you a hint of the horizon line to settle your stomach. And the gunsman sleeping in the hammock next to you has dysentery. That was the reality of life on a ship like the *Charles II*.

As horrendous as the sleeping conditions were, they paled in comparison to the food. If they ever build a library of literary disgust, a special wing will no doubt be dedicated to sailors describing the atrocities that they were forced to consume at sea. A few decades after the *Charles II* set sail for the West Indies, the captain of another privateering mission recounted the horrifying conditions on board as the ship slowly ran through its supplies of food and drink: "We constantly drank our urine, which, though it moisten'd our mouths for a time, excited our thirst the more . . . [O]ur common food was puddings made of very coarse flour and sweetmeats, and salt water instead of fresh to moisten them, and dry'd beef, which was partly destroy'd by ants, cockroaches, and other vermin." Another privateer from the late 1600s described the Christmas dinner the crew enjoyed on one voyage: "For we had nothing but a little bit of Irish beef for four men, which had lain in pickle two or three years and was as crusty as the Devil, with a little stinking oil or butter, which was all the colours of the rainbow, many men in England greasing their cartwheels with better."

Then there was the question of where all that food was supposed to go once it had been eaten. The primary toilet on board was a hole suspended over the bowsprit at the head of the ship. (The modern slang for going to the toilet—"hitting the head"—originated on these vessels.) Needless to say, the close quarters and the nonexistent sanitation—not to mention the travels to exotic lands with parasites and microorganisms that Europeans had never encountered before— meant that the crews aboard these ships often faced catastrophic medical situations. Hollywood representations of pirates and privateers tend to focus on the battle scenes, with cannonballs firing and elaborate swordfights on deck, but the reality of life at sea during this period was that you were more likely to die from the "bloody flux"—as dysentery was called back then—than you were to be

struck down in armed combat. While there was some vague aware-
ness by Every's time that the excruciating—and often fatal—vitamin
C deficiency known as scurvy could be treated by adding citrus to
the diet, the condition was nonetheless rampant on commercial and
military vessels during the period. Sexually transmitted diseases,
too, accompanied the seafaring lifestyle, thanks both to encounters
with prostitutes in ports of call and to onboard encounters among
the sailors themselves. One study of thirty-three Royal Navy ships
found that almost 10 percent of the crew suffered from venereal diseases of one sort or another—on some ships, one in four men had
contracted them—in an era where now treatable diseases like syphilis were often fatal.

With their cramped conditions and limited food supplies, the
ships were petri dishes for disease and malnourishment—in both
senses of the term: the ships' conditions made them a breeding
ground for potential pathogens as well as a tool of scientific research.
Ships like the *Charles II* afforded the medical men aboard tightly controlled experiments: a hundred men sharing almost the exact same
circumstances for long periods of time. It is no accident that what
many consider to be the first controlled clinical trial in history took
place on board one of these vessels, when the Scottish doctor James
Lind experimented in 1747 by giving members of his crew different
remedies for scurvy—among them cider, sulfuric acid, seawater, and
oranges—and carefully observing the results.

Not all ship doctors were as empirical as Lind. The default cure
for most illnesses was still bloodletting. The "bloody flux" was
treated with a now comical range of interventions, everything from
sitting on a heated brick to being buried up to the neck in hot sand to
an unorthodox variation on a modern suppository: "Take a hard
egg," one medical guidebook advised, "and peel off the shell, and

put the smaller end of it into the fundament or arsehole, and when that is cold take another such hot, fresh, hard, and peeled egg and apply it as aforesaid."

The atrocious medical conditions and limited access to proper nutrition meant that the mortality rates for voyages like the Spanish Expedition—even if you don't count the ventures that ended in ship-wreck and a complete loss of life—were appallingly high, in an age where the average life expectancy was just over thirty years. According to a mate on William Dampier's 1706 voyage around the world, Dampier began his journey with a 183 men under his command. By the time he returned, "after many Dangers both by Sea and Land," only eighteen remained.

So if you want to put yourself into the mind of Henry Every, suffering through those long months off the coast of Europe, waiting to be set free to begin the true journey of the Spanish Expedition, you should imagine this: you are living in a floating aquatic coffin with a hundred other sailors, just trying to carve out a few feet of space to sleep every night without contracting smallpox from the men rubbing shoulders with you as you sleep.

But consider one other thing when you imagine yourself in Henry Every's situation. Imagine *choosing* to live your life this way. Because for most of the men, the life at sea was a deliberate choice. For every William Bishop, the teenager who claimed to have been forced into service on the Spanish Expedition, there were probably at least ten Everys or Dampiers, who actively sought out privateering as a career path. The conditions they lived in were far more physically taxing and oppressive—not to mention deadly—than the living conditions of just about any human being currently alive in the twenty-first century, even in the poorest or most remote communities on the planet.

Why did they sign up in the first place? The wages alone were a temptation. In those alehouses and taverns below London Bridge, an informal market had emerged where able seamen could offer their services to multiple employers—one of the earliest professions to generate that kind of competitive marketplace for labor. (A little more than half a century after the Spanish Expedition left London, sailors would stage one of the first general strikes in labor history. The word "strike" itself derives from their strategy of "striking," or lowering, the sails of anchored ships as a sign of their refusal to work.) An able seaman signing up for a voyage like the Spanish Expedition could attract basic wages comparable to those of skilled craftsmen like tailors and weavers. And while the provisions were often contaminated with weevils and ants, the sailors were guaranteed food and drink for the length of the voyage as part of their contract—a significant savings in an era where most land-based workers spent almost all their wages warding off starvation.

There was genuine camaraderie aboard the ship, as well. The sailors entertained themselves with card games and music. Literacy rates were surprisingly high: according to one study, more than 70 percent of sailors could sign their own names on official documents, the standard used by demographic historians to gauge literacy rates in earlier societies. An amateur science and travel writer like William Dampier was hardly the norm, of course, but reading books and pamphlets was an available pastime to many on board. Sexual experiences, too, were part of the attraction: the fantasy of a sexualized "Orient," liberated from the religious strictures of Christian Europe. Port Royal in southeast Jamaica—for a time the largest city in the entire Caribbean—was known as the "Sodom of the New World." Not all the sexual contact took place in harborside brothels. While both formal and informal legal codes explicitly forbade sodomy, on some ships meaningful same-sex relationships appear to have been condoned.

A life at sea also offered access to radically new experiences, something that was otherwise in short supply in the seventeenth century, in both Europe and India, despite their relative affluence. Growing up working class in Devonshire in the 1660s, Henry Every would have had almost no opportunity to encounter other realities beyond the insular world of a fishing town on the English Channel. Novels were still in their infancy; Dampier's groundbreaking travelogues—which offered a simulated ride on a globetrotting voyage—wouldn't be published for another forty years. The theater could immerse you in a virtual world, and the church built spaces designed explicitly to dazzle the senses, but the spells of both those spaces were limited, particularly to lower-income people who weren't attending plays in fashionable London. Travel, on the other hand, was the real thing. If you genuinely wanted to expand your horizons, you had to do it the old-fashioned way.

All these attractions—the financial incentives, the sexual and geographic adventure of the privateering lifestyle, the whole romance of the sea—were to a certain extent amplified by one of the new inventions of the seventeenth century: the popular press. The early pioneers of print media quickly discovered that there was a paucity of recognizable people whom they could cover in their broadsides and pamphlets. The average British citizen circa 1500 would have heard almost nothing about other human beings outside his or her immediate network of friends, family, and neighbors. The only living people with genuine national name recognition were members of the royal family and some lesser orbit of political figures—along with the upper echelons of the clergy. There were no rock stars or billionaire entrepreneurs or reality-television stars who were properly famous. The printing press had created the possibility—for the first time in human history—of assembling a mass audience, but that audience didn't have enough social common denominators to be a meaningful (or at least profitable) unit. The exotic and scandalous lives of privateers supplied one of the first solutions to that problem. Figures like Francis Drake became legends in their own time, men from humble beginnings who had sailed their way into affluence and prestige. In this sense, pirates and privateers were forerunners of the modern celebrity. Henry Every would achieve notoriety as the most wanted man in the world, but from a historical perspective, the *scale* of that notoriety was just as noteworthy: few human beings had ever captured the imagination of so many strangers around the world without commanding an army, presiding over a major religious sect, or being born with royal blood.

Of course, one of the reasons Every became so famous is the simple fact that his actions made him, for a time, astonishingly wealthy, at least by the standards of a humble Devonshire seaman. Even before they had Every as a role model, the generation of sailors signing up for service on a privateering mission like the Spanish Expedition had the promise of outsize reward to lure them to sea. A venture like the Spanish Expedition could pay out the equivalent of ten years of wages to a first or second mate like Every or Dampier if it got lucky, even after sharing a bounty with the investors back in London. In an age where class mobility was for all intents and purposes nonexistent, heading out to sea in search of treasure was the one viable path to changing your station. All the risk of disease and shipwreck and starvation were worth the potential reward, given the limited options at home.

But the longer the men of Spanish Expedition Shipping waited at the Groyne, the less likely that financial upside became. Back in London, several wives of the crew had tracked down James Houblon and personally demanded the back pay that was owed to the families according to the contracts their husbands had signed. Houblon icily responded that the men belong to the king of Spain now, and were no longer his concern. Instead of seeking their fortune in the West Indies, the crew were trapped in a bureaucratic holding pattern, without even the solace of reliable wages. When Houblon's dismissive remarks traveled back to A Coruña, rumors began to swirl that they were all going to be sold into slavery to the Spanish.

As the crew passed the days and nights in the taverns of A Coruña, increasingly convinced that the Spanish Expedition would never make it to the West Indies, a new plan began to form in the mind of the *Charles II*'s first mate.

Part Two

THE

MUTINY

9

THE DRUNKEN
BOATSWAIN

A Coruña

May 7, 1694

I n the soft glow of late twilight, with a quarter moon overhead, the medieval fort of A Coruña is still visible as the longboat pulls quietly up against the *James*.

On the main deck of the *James*, First Mate Thomas Druit is on watch. From the longboat, a voice calls, "Is the drunken boatswain on board?" Confused by the question, and unable to make out the speaker's face in the low light, Druit gives a quizzical reply. The stranger on the longboat mutters a brief warning—the *Charles* "is going to be run away with"—and then pushes off into the black water.

If the "drunken boatswain" question is baffling to First Mate Druit, its meaning is all too apparent for other members of the crew. The past few weeks, over drinks in the taverns in A Coruña and in hushed conversations belowdecks on the *Charles II*, Henry Every and a handful of sailors have been plotting to stage a mutiny aboard the

Charles. They have settled on "the drunken boatswain" as a password, a signal that the uprising is under way, at long last liberating the men from their five-month holding pattern in A Coruña's harbor.

A few hundred feet away, on board the *Charles*, Second Mate David Creagh makes his way across the quarterdeck to check on Captain Gibson, who has fallen ill, likely with some combination of fever and alcohol poisoning. Before reaching the captain's quarters, Creagh encounters a group of men—including the middle-aged sailor William May, Henry Every, and the ship's carpenter—enjoying a bowl of punch. After a quick visit with Gibson, Creagh sits down with the sailors. They seem to be in unusually good cheer. May proposes a toast: "A drink to the health of the Captain, and the Prosperity of our Voyage." It is an odd tribute given the grim prospects currently facing the Spanish Expedition, spoken by a man who might well be about to be sold into slavery to the Spanish king. But Creagh raises his glass anyway, and then heads down to his bunk.

Back on the *James*, Druit alerts Captain Humphries about the ominous warning from the longboat. Humphries immediately commands Druit to "man the pinnace," the longboat attached to the *James*, generally used to ferry crew and goods to and from the harbor. Eighteen-year-old William Bishop dutifully follows orders and climbs down into the longboat, preparing to rescue the *Charles* from its usurpers. But almost immediately he finds himself surrounded by fifteen men who have another plan in mind. The enigmatic question about the "drunken boatswain" has succeeded in transmitting its secret message. Before Druit can get back to the pinnace, twenty-five men have already boarded her, among them Edward Forsyth, James Lewis, and young Bishop. Druit orders them to return, but they ignore his command and row furiously toward the *Charles*.

With evidence of the mutiny beyond dispute, and the already troubled Spanish Expedition now in serious peril, Captain Hum phries faces a terrible quandary: Should he let his own men go freely to capture the *Charles*, with potentially deadly consequences for his peer, Captain Gibson? Or should he take an equally devastating step and fire on his own crew?

Every's supporters gather on the deck of the *Charles* as their fellow mutineers from the *James* row across the harbor toward them. The plan had originally been to confront Gibson while still anchored in A Coruña, once all the rebels had reunited on board, presenting him with an offer of a bloodless transfer of power.

Two shots across the bow from Humphries and the *James* scuttle that plan in a matter of seconds. Whatever negotiations are needed with Captain Gibson will have to happen on the open sea.

The mutineers on the pinnace are quickly hauled aboard. Joseph Gravet, a second mate of the *Charles*—presumed to be loyal to Captain Gibson—is seized, a "pistol clapt to his breast," and put under armed guard belowdecks. Henry Every takes control of the *Charles*; the anchor lines are cut and sails unfurled. Every orders that the pinnace from the *James* be set adrift. Somehow Druit and Humphries are able to alert the A Coruña fort about the mutiny in progress, and the *Charles* sails out of the harbor chased by cannon fire from both sea and land. Not for the last time, the ship's legendary speed in the water proves to be its salvation.

THE MUTINY ABOARD the *Charles II* is one of those rare moments from history where we can re-create an almost second-by-second account of the actions—down to the words the participants used with each other as the event was unfolding—despite the fact that the whole affair took place more than three hundred years ago, and hardly suggested that it would lead to world-shaking consequences at the time.

One of the ironies of the mutiny and its historical legacy is that the man who would become the single most influential and accomplished nautical author of his age was aboard the *Dove* as the events were happening, in the form of William Dampier. And yet none of our second-by-second account of the mutiny comes from his writing. Dampier never published a word about the events of May 7, 1694. The mutiny was preserved by a court reporter in London, years later, transcribing sworn testimony from the participants. By that point the historical significance of the mutiny would be plain to see.

LYING IN HIS HAMMOCK in the forecastle, Creagh springs up at the sound of the cannon fire. His punch-bowl suspicions have been confirmed: something out of the ordinary is clearly unfolding. He clambers his way up to the quarterdeck, where he finds Every at the helm, steering the ship out of the harbor. Beside him stands the ship's carpenter, seemingly now playing the role of Every's muscle.

Every grabs Creagh by the hand, and asks, "Will you go with me?"

As Every stares him down, Creagh equivocates, saying, "I do not know your design."

After a tense and cryptic exchange, Every tells him, "You will all know by tomorrow morning, eight o'clock."

The carpenter intervenes, pulling Creagh aside and pointing fiercely in Every's direction. The carpenter's words—preserved by Creagh's testimony years later—have a bawdy, Shakespearean ring to them.

"Do you not see this Cock?" the carpenter demands.

Creagh nods. "I do."

"This man—and old May and Knight—I can trust with anything. They are true cocks of the game, and old sports-men." Then he delivers the threat: "If you do not go down I will knock you on the head."

Fearing for his life, Creagh heads back belowdecks. As he descends, he finds William May near the hatch. According to Creagh, the old sea dog challenges him: "What do you do here?" Trying to stay out of the conflict erupting around him, Creagh ignores the question and continues on toward his cabin. May presses a pistol against his skull and offers a curse, one that would return to haunt the aging sailor years later at the trial: "God-damn you. You deserve to be shot through the head."

CREAGH IS NOT the only one alarmed by the cannon fire. In the captain's quarters, Gibson finally rouses from his fever dream. He can feel the ship fighting through the swell of the open Atlantic. Stumbling out onto the quarterdeck, he finds himself face-to-face with Every.

Gibson struggles to make sense of the situation. "Something is the matter with the ship. Does she drive? What weather is it?"

"No, no we're at sea, with a fair wind and good weather."

Even more confused, Gibson exclaims, "At sea! How can that be?"

Every lays out the situation. "I am captain of this ship now, and this is my cabin," he says. "Therefore you must walk out. I am bound to Madagascar, with a design of making my own fortune, and that of all the brave fellows joined with me."

At this point Henry Every offers the captain a deal, the exact details of which are a matter of some dispute. By some accounts, Every proposes to swap rank, with Gibson serving as first mate under the newly appointed Captain Every. But according to Creagh's version of the story, Every makes an even more generous proposition, saying, "If you will go in the ship, you shall still command her."

Gibson rejects Every's offer out of hand. "I never thought you should have served me so, who have been kind to all of you," he stammers. "And to go on a design against my owner's orders—I will not do it."

Every nods. "Then you must go ashore."

The historical record is clear on one thing: Every and his backers allow Gibson to make an honorable exit from the *Charles*. (Not all mutinies ended with such a civil separation.) After his session with Gibson, Every pays a visit to Second Mate Gravet, under house arrest in his cabin.

"I suppose you do not intend to go with us," Every says.

When Gravet confirms Every's hunch, the new captain of the *Charles* extends the same amnesty to the rest of the crew that he has just extended to Gibson: "You, and the rest that will not go with us, have the liberty to go ashore." But their departure must be immediate. Gravet is to be led up to the longboat "carrying no more than the clothes on [his] back."

Every's apparently genuine desire to see the former captain and second mate ashore is complicated by one fact. By first light, when the newly constituted crew of the *Charles*—some mutineers, some still loyal to James Houblon and the Spanish Expedition—gather on deck to assess their conditions, the *Charles* is ten miles out to sea. Gibson and his supporters will have to row back to shore in the *Charles*'s pinnace.

In one of his final moments onboard the *Charles*, Gravet passes William May, who takes him by the hand and wishes him well. According to Gravet, May is "very merry and jocund" as they part. His final words to Gravet are: "Remember me to my wife." At the last minute, Every has a few clothes brought up for the second mate: a coast and waistcoat, along with the letter of rank (otherwise known as a commission) that he had left behind belowdecks. Their parting words are amicable. There is no known account of Every's last exchange with the deposed Captain Gibson.

Within minutes seventeen men are in the longboat, preparing to row back to A Coruña, whose fortress has long ago disappeared beneath the horizon. As they push off, they notice that their lifeboat is taking on water at an alarming rate. As experienced seamen, they can do the calculations with the speed of instinct: a boat leaking this badly will not make it ten miles back to shore. Drifting away from the *Charles*, Gravet and his mates scream for a bucket to be tossed to them. For a moment it looks as though all the chivalry of the bloodless mutiny is about to be exposed as a fraud: instead of being shot in the head, they are to drown at sea. But the crew on board the *Charles* toss a bucket to the pinnace, and the seventeen faithful servants of the Spanish Expedition begin the long slog back to A Coruña.

Captain Gibson must have noticed, as he surveyed the sixteen men who had stayed loyal to him, that the pinnace of the *Charles* still had room for many more.

THOSE EMPTY SEATS on the pinnace would ultimately be invoked as clear evidence of criminal intent among the crew remaining on board the *Charles*. Two facts are undisputed: Every allowed a significant portion to leave in peace, and there was room for more on the longboat headed back to harbor. Viewed together, they made a damning case against the mutineers: the men had been offered a chance to leave, but they had chosen to stay of their own volition.

And yet despite the consensus around those two facts, and despite all the details that have survived about the events of May 7, 1694, there remains a *Rashomon*-like quality to the mutiny. The same events—even the same words—take on different meaning if you entertain the premise that some of the mutineers were forced to remain with Every against their will. In one telling of the story, William May comes across as a key ally of Every's: merrily proposing a toast to their new mission; threatening to shoot Creagh through the head for not joining the revolt; parting happily with Gravet as the former second mate boards the longboat. But May would claim for the rest of his life that he had not been a willing participant in the mutiny, that his toast at the punch bowl had held no ulterior meaning, that his threat to shoot Creagh had been act of loyalty to Gibson, not Every.

In May's own account of the mutiny, shortly after his confrontation with Creagh, he approaches Every at the helm of the *Charles*. Every senses that May is not likely to be loyal to him. "You, May, I believe you do not love this way," he says. "Pray get down to your Cabin." He descends back to his quarters and weighs his options. In his testimony, May described what happened next:

> I was thinking: I must [not] leave me old Captain without seeing him; I begged them to give me leave to come to him; and there was two men stood with naked Cutlaces, and would not let me come to him. We had some confabulation together, and I begged the favor to come in, and at last they permitted me; and the Doctor was anointing the Commander's temples. . . . When I came out again, they began to hurry the men away. Here was Mr. Gravet, the Second Mate . . . I told him he should remember me to my Wife, as I am not likely to see her, for none could go, but who they pleased.

None could go, but who they pleased. Rewrite the narrative along these lines, and May's final words to Gravet take on an entirely different valence: a man asking mournfully to be remembered to his wife not because he is choosing to set sail with a band of mutineers, but rather because he has no choice, because he knows his odds of return have just become far less promising. In this version, May is not Every's deputy and co-conspirator, but rather his prisoner.

They were just words, no more than five or six sentences: a drunken toast to a ship's captain, a threat leveled in a hatchway, a message to be passed back to a spouse. But how you read those simple statements—the inflection you give them—turned out to be the difference between life and death for William May.

IF MAY—and others on board the *Charles*—were not willing participants in the mutiny, why then did they not fill those empty seats and join Gibson and Gravet on the journey back to the *James*? According to May, the issue was the seaworthiness of the longboat itself. "When those men were in the boat," May would later testify, "they cried for a bucket or else they should sink, they having three leagues to go. And I did not know how they could go so far with more, when their boat was likely to sink with those that were in her." In May's account, he had been given a false choice: he could enter a life outside the law and become a mutineer, or he could drown with seventeen men on the open waters of the Atlantic. "If I should have denied to go with them, I might have been killed by them," May would argue in the courtroom. "And I knew not whether it be better to be accessary to my own death, or to suffer by the Law of the Nation."

In the end, roughly eighty men would set sail with the *Charles*, and renounce their ties to Spanish Expedition Shipping. As they gained speed, Every rechristened the vessel. The men were sailing on the *Fancy* now—an allusion both to the quality of the ship, and the treasure they hoped to secure with her. The men, too, had been given a new name. Whatever their original role had been in the mutiny—ringleader, henchman, captive—they were all pirates now.

10

THE *FANCY*

The Atlantic Ocean, west of Africa
May–June 1694

One of Henry Every's first actions as captain of the *Fancy*—now that he and his crew had severed ties irrevocably with James Houblon and the Spanish Expedition—was to establish what we would now call a profit-sharing plan for their enterprise. While pirates have almost always lived outside the laws of nation-states, and while they have a sometimes deserved reputation for anarchic acts of violence, within their floating communities they usually created—and obeyed—surprisingly coherent codes that governed their behavior, including their financial interactions. Most pirate voyages began by establishing "articles of agreement," the bylaws that would shape both the political and economic relations among captains, officers, and ordinary crew.

The most critical article involved the distribution of loot. Much like the investors in the East India Company, each pirate was considered a shareholder in the venture. If they were lucky enough to seize treasure during their voyage, the bounty would be distributed based on the shares held by each man. But unlike the East India

Company—and indeed just about any modern corporation—the distribution of profit on almost all pirate ships was radically egalitarian. To give some frame of reference, the compensation for American corporate executives today is, on average, 271 times larger than the median worker compensation in the firm. On a Royal Navy ship during Every's time, the captain and officer class might earn ten times the wages of the average able seaman. On a merchant vessel, or a privateering mission like the Spanish Expedition, the income ratio could be as low as five to one. Pirate distributions were even flatter. The articles on board eighteenth-century pirate Edward Low's ship—named the *Fancy* in honor of Every—spelled out the economic terms as follows: "The Captain is to have two full shares; the Master is to have one Share and one half; The Doctor, Mate, Gunner and Boatswain, one Share and one Quarter." The rest of the crew were granted one share a piece. Henry Every and his men adopted a simpler structure: two shares for Every, one share for everyone else.

If the other articles of agreement aboard the *Fancy* were written down at the time, they have not survived in the historical record. But we do have four complete articles of agreement from pirates doing business in the decades after Every's voyage: Low, Bartholomew Roberts, John Phillips, and George Lowther. They are fascinating documents, in the glimpse they give of both the everyday pastimes on board a pirate ship, and the surprisingly nuanced political systems the pirates developed to maintain order and secure stable governance on their voyages. Of the four surviving articles, Roberts's makes for the most vivid reading:

> I. Every man shall have an equal vote in affairs of moment. He shall have an equal title to the fresh provisions or strong liquors at any time seized, and shall use them at

pleasure unless a scarcity may make it necessary for the common good that a retrenchment may be voted.

II. Every man shall be called fairly in turn by the list on board of prizes, because over and above their proper share, they are allowed a shift of clothes. But if they defraud the company to the value of even one dollar in plate, jewels or money, they shall be marooned. If any man rob another he shall have his nose and ears slit, and be put ashore where he shall be sure to encounter hardships.

III. None shall game for money either with dice or cards.

IV. The lights and candles should be put out at eight at night, and if any of the crew desire to drink after that hour they shall sit upon the open deck without lights.

V. Each man shall keep his piece, cutlass and pistols at all times clean and ready for action.

VI. No boy or woman to be allowed amongst them. If any man shall be found seducing any of the latter sex and carrying her to sea in disguise he shall suffer death.

VII. He that shall desert the ship or his quarters in time of battle shall be punished by death or marooning.

VIII. None shall strike another on board the ship, but every man's quarrel shall be ended on shore by sword or pistol in this manner. At the word of command from the quartermaster, each man being previously placed back to back, shall turn and fire immediately. If any man do not, the quartermaster shall knock the piece out of his hand. If both miss their aim they shall take to their cutlasses, and he that draweth first blood shall be declared the victor.

IX. No man shall talk of breaking up their way of living till each has a share of 1,000. Every man who shall become a cripple or lose a limb in the service shall have 800 pieces of eight from the common stock and for lesser hurts proportionately.

X. The captain and the quartermaster shall each receive two shares of a prize, the master gunner and boatswain, one and one half shares, all other officers one and one quarter, and private gentlemen of fortune one share each.

XI. The musicians shall have rest on the Sabbath Day only by right. On all other days by favour only.

Some of the principles in this mini constitution—composed sometime in the 1720s—seem appropriately archaic to the modern mind: most political documents today do not specify the terms for dueling, or forbid candlelight after eight p.m. But on the most important points, the pirate codes—as the articles of agreements were sometimes called—were significantly ahead of their time. Consider the opening line of the Roberts articles: "Every man shall have an equal vote in the affairs of moment." The pirates encoded these democratic principles into their constitutions almost a century before the American and French Revolutions. A captain served at the pleasure of his crew, and could be removed from power if he fell out of favor with the majority. Navy and merchant ships were autocratic institutions, with a tightly controlled chain of command headed by a captain possessing absolute authority over the ship, and no mechanism for curbing any abuse of that power. The pirate ship, by contrast, was a floating democracy. According to Charles Johnson's 1724 bestseller, *A General History of the Pyrates*, which included a long chapter on Every and his crimes, on a pirate ship "the supream Power lodged with the Community, who might doubtless depute and revoke as suited Interest or Humour."

The elegance of the pirate governance model went beyond their voting rights. Most pirate ships during the period created a separation of powers on board that bears a striking resemblance to the architecture of the US Constitution. The captain's authority was not just limited by the threat of being voted out of office; he was also reined in by the separate authority of the quartermaster. While the captain had unrestricted powers during battle, and had executive authority at all times to establish the overall mission for the ship, most day-to-day issues were adjudicated by the quartermaster, who also was charged with the distribution of prizes. In Johnson's account:

> For the Punishment of small Offences . . . there is a principal Officer among the Pyrates, called the Quarter-Master, of the Men's own choosing, who claims all Authority this Way, (excepting in Time of Battle:) If they disobey his Command, are quarrelsome and mutinous with one another, misuse Prisoners, plunder beyond his Order, and in particular, if they be negligent of their Arms, which he musters at Discretion, he punishes at his own dare without incurring the Lash from all the Ship's Company: In short, this Officer is Trustee for the whole, is the first on board any Prize, separating for the Company's Use, what he pleases . . .

If the captain served as an elected leader, roughly equivalent to a president or CEO, the quartermaster played a more eclectic role, a mix of a judicial branch, determining punishments for onboard transgressions, and a chief financial officer, overseeing compensation packages. The quartermaster's authority meant that he was often first in line to succeed a deposed captain, not unlike the transfer of power that brought First Mate Every to the helm of the *Charles II* during the mutiny. In those first few days after leaving A Coruña, the crew appointed a quartermaster to serve alongside Every. His name was Joseph Dawson, a veteran sailor in his late thirties, originally from Yarmouth.

The crew of the *Fancy* would have likely also established terms for another key innovation that pirate codes introduced, one that can be seen in Article IX of Roberts's pirate code: "Every man who shall become a cripple or lose a limb in the service shall have 800 pieces of eight from the common stock and for lesser hurts proportionately." Pirate communities built insurance into their constitution as a core principle of the collective. Pirates who suffered serious injuries in battle would receive a disproportionate share of whatever treasure the group managed to secure. Some insurance schemes were more precise than the one outlined in the Roberts articles. According to eighteenth-century pirate and slave trader Alexandre Exquemelin, wounded crewmen were offered specific levels of compensation depending on the injury: losing a right arm was worth slightly more than losing a left arm; losing an eye generated almost the same compensation as the loss of a finger.

All these elements combined—an onboard democracy, with separation of powers; equitable compensation plans; insurance policies in the event of catastrophic injuries—meant that a pirate ship in the late 1600s and early 1700s operated both outside the law of European nation-states and, in a real sense, *ahead* of those laws. The pirates were vanguards as much as they were outlaws, building codes that ensured the collective strength of the ship and guarded against excessive concentration of both power and wealth. At the very moment the modern multinational corporation was being invented, the pirates were experimenting with a different kind of economic structure, closer to a worker's collective. Those economic and governance codes have led historians in recent years to reevaluate the place of the pirates, seeing them now not just as significant figures in the history of crime and exploration, but also as pioneers in the history of radical politics. As the maritime historian J. S. Bromley wrote, the pirates "were not merely escaping from bondage. In their enterprises at least, they practiced notions of liberty and equality, even of fraternity, which for most inhabitants of the old world and the new remained frustrated dreams, so far as they were dreamt at all."

In his magisterial account of seafaring culture in the early 1700s, Marcus Rediker writes of the politics of the pirate class:

> Pirates constructed a culture of masterless men. They were as far removed from traditional authority as any men could be in the early eighteenth century. Beyond the church, beyond the family, beyond disciplinary labor, and using the sea to distance themselves from the powers of the state, they carried out a strange experiment . . . [The pirates] expressed the collectivistic ethos of life at sea by the egalitarian and comradely distribution of life chances, the refusal to grant privilege or exemption from danger, and the just allocation of shares.

Understanding that egalitarian ethos is essential to understanding why pirates like Henry Every were so popular at home. They were not just charismatic rogues, pursuing a life of adventure at sea. They were also advancing populist values that had almost no equivalent on the mainland.

Of course, for that working-class-hero myth to take hold, the pirates needed more than mere word of mouth to get the message out. They also needed the amplifying power of media. Henry Every and the crew of the *Fancy* would soon enough become genuine celebrities, their exploits glorified and condemned in just about every form of media in existence at the end of the seventeenth century, in pamphlets, book-length biographies, published transcripts from criminal trials, even dramatic plays. But word of Henry Every's turn to piracy would first captivate working-class audiences back home through a far older medium: song.

11

THE PIRATE VERSES

London
June 1694

I t is not known when exactly James Houblon and the other investors behind the Spanish Expedition learned that Henry Every and
his men had seized the *Charles* and set off to seek their fortune as
pirates. As strange as it might seem to us today, the first surviving
notice of the mutiny back in London is neither a newspaper item nor
a legal complaint nor business correspondence. Instead, the earliest
record of Every's betrayal appearing in England is a poem.

Sometime in the late spring or early summer of 1694, a London
printer named Theophilus Lewis published a thirteen-stanza ballad
with the title "A copy of Verses, composed by Captain Henry Every,
lately gone to Sea to seek his Fortune." As enticing as it is to imagine
the newly appointed Captain Every tinkering with his rhyme
schemes as he sails past the Strait of Gibraltar, the verses published
by Lewis were almost certainly written by someone other than
Every himself. Multiple versions of the "Every Verses," as they

became known, would be published over the next few years, and the slight variations between them suggest that they are all derivations of an earlier rendition of the poem, likely transmitted orally as a sung ballad.

Like so many events in Henry Every's life, the first reports of his mutiny aboard the *Charles II* emerged at a critical transition point between two distinct regimes: in this case, the transition between song and print. Throughout the 1600s in major European cities like London, military and political news, folklore, and true-crime narratives were transmitted via songs, in a tradition that descended from the minstrel singers of the pre-Gutenberg era. But as printing technology expanded, those sung ballads were increasingly accompanied by paper versions, with text and woodcut illustrations printed on one side of a large sheet of paper. Popularly known as a broadsides, these musical accounts of current events would be hawked by balladmongers on street corners. Modern newspapers would ultimately evolve out of these early experiments in printed news, and the balladmongers themselves were the antecedent of the classic newsboy that survived all the way into the twentieth century, hawking papers by working headlines into his standard refrains: "*Titanic* sinks! Read all about it!" But the balladmongers plied a more musical craft: they would actually sing the news to attract buyers for their latest broadsides, setting the "headlines" to a recognizable tune. The Every Verses, for instance, were set to the tune of a then familiar song called "The Two English Travellers."

Imagine strolling in early June 1694 through the neighborhoods at the foot of London Bridge—Limehouse and Wapping and Rotherhithe—their alehouses crowded with sailors and merchant ship representatives pitching new ventures along the lines of the Spanish Expedition. Above the dissonant clamor of horse hooves on cobblestone, the bellowing pitches of the street vendors, the slurred arguments breaking out in front of each pub, rises the crooning of the balladeers, as they wave their broadsides at you. Their lyrics tell tales of political intrigue or unusual weather or gruesome murders: all the current events that now lead the eleven o'clock news, sung to you as you wander through the streets of the big city. In the late spring of 1694, you might have heard a ballad called "The Loyal British Fighting in Flanders," allegedly composed by a "Protestant Sentinel of the British Forces." Or you might have whistled along to a darker verse, one called "The Murtherers Lamentation." (The explanatory copy on the broadside reads: "Being An Account of *John Jewster* and *William Butler*, who where arraign'd and found guilty of the Robbery and Murder of Mrs. *Jane Le-grand*; for which they received due Sentence of Death, and was accordingly executed. . . .") But on one corner, a balladeer serenades you with a new set of verses. Technically, it tells the story of a crime, with the promise of more crimes to come, but its rhetorical form is a call to action, an entreaty:

> Come all you brave Boys, whose Courage is bold,
> Will you venture with me, I'll glut you with Gold?
> Make haste unto Corona, a Ship you will find,
> That's called the *Fancy*, will pleasure your mind.
>
> Captain Every is in her, and calls her his own;
> He will box her about, Boys, before he has done:
> French, Spaniard and Portuguese, the Heathen likewise,
> He has made a War with them until that he dies.
>
> Her Model's like Wax, and she sails like the Wind,
> She is rigged and fitted and curiously trimm'd,

And all things convenient has for his design;
God bless his poor *Fancy*, she's bound for the Mine.

The ballad contains enough key facts about the mutiny on the *Charles II*—facts that had not been reported elsewhere in London at the time—to suggest that it must have been written by someone with first- or secondhand knowledge of the events of early May. A later line alludes to Every's Devonshire roots, but suggests a (likely untrue) connection to a prominent landed family in that region. ("Farewel, fair Plimouth, and Cat-down be damn'd / I once was Part-owner of most of that Land; But as I am disown'd, so I'll abdicate / My Person from England to attend on my Fate.") The verses also accurately describe Every's itinerary:

Then away from this Climate and temperate Zone,
To one that's more torrid, you'll hear I am gone,
With an hundred and fifty brave Sparks of this Age,
Who are fully resolved their Foes to engage.

These Northern Parts are not thrifty for me,
I'll rise the Anterhise, that some Men shall see
I am not afraid to let the World know,
That to the South-Seas and to Persia I'll go.

Beyond the factual account of the *Fancy*'s mission, the verses perform a kind of "song of myself," announcing Every's ambition, the same design to seek his fortune that he declared to Gibson on board the *Charles* on May 7. You can hear in its couplets the first stirrings of a narrative device that would animate a thousand maritime tales of self-made men—the now familiar Horatio Hornblower narrative, of an ambitious sailor who pulls himself up by his bootstraps and becomes a legend:

> Our Names shall be blazed and spread in the Sky,
> And many brave Places I hope to descry,
> Where never a French man e'er yet has been,
> Nor any proud Dutch man can say he has seen.

> My Commission is large, and I made it my self,
> And the Capston shall stretch it full larger by half;
> It was dated in Corona, believe it, my Friend,
> From the Year Ninety three, unto the World's end.

As the historian Joel Baer notes, the Every Verses are interesting in the lack of attention paid to the actual details of the mutiny itself. An in-depth account of how Every came to seize the ship would have been an easy fit with the true-crime narrative that so many of the balladmongers hawked. "A song of kindly commanders, ambitious tars and brutal ingratitude might have been fashioned that conformed to the conventions of the news ballad," Baer observes. "The writer chose rather to shift the focus from the overt act of disloyalty and theft to the narrator's convictions about himself, his crew, and the society he is about to abandon." The Every Verses also broke from convention in another key respect. With the notable exception of the Robin Hood ballads, almost all verse-based crime reporting in the period told stories of crime retrospectively, from the jail cell or the gallows. (Broadsides were often sold at the scenes of public executions, like a libretto handed out at an opera.) Audiences were encouraged to thrill and shudder at the tales of murder and robbery, but the moral boundaries that framed these stories were clearly defined: the criminals deserved the punishment they inevitably received. In the Every Verses, Henry Every is never painted as a criminal, despite the obvious facts of his crime; he's an inspirational figure, with a stirring message for all the "brave boys whose courage is bold."

The Every Verses are significant for another reason. For the first time in Every's life, a second front had opened up in his identity: There was Every the sailor, headed south with his crew toward the Cape of Good Hope on board the *Fancy*. And then there was a second Every: the one the balladmongers were celebrating on the streets of London, who would develop his own mythology over time as his story moved from sung ballads to broadsides to books and the stage, a mythology that often deviated significantly from the actual course of his life. For the generation of pirates that followed in his wake—as well as for the nation of England itself—the mythical Every would prove to be just as influential as the real thing.

12

DOES SIR JOSIAH SELL OR BUY?

London
August 1694

The "Every Verses" had originally been intended as an amusement, designed to delight London consumers with the ballad of "Bold Captain Every." But by the end of the summer of 1694, they would become evidence in a legal dispute. In June, just as the "Every Verses" were beginning to be sung on the street corners of London, the aggrieved wives of the remaining Spanish Expedition crew petitioned the Crown to intervene in their dispute with James Houblon. The Spanish Expedition investors—led by Houblon himself—had behaved "traitorously," they claimed, leaving their husbands "in to the King of Spaines Service to Serve him, as far as [we] know all the dates of their lives." Before long, the Privy Council had opened an investigation, inviting Houblon to submit evidence in his defense. Houblon produced three documents: a list of the Spanish Expedition investors, presumably designed to impress the council with the investors' social stature; the employment contracts signed by the crew; and a broadside copy of "Verses, composed by Captain Henry Every, lately gone to Sea to seek his Fortune."

While it seems unlikely that Every himself composed the ballad, Houblon took the document at its word. In his submission to the Privy Council, he observed that the Every Verses were a "Declaration of their intentions of Pyrating, Greatly to the Dishonour of this Nation and damage to the Owners' that the mutineers had left behind them at La Coruna."

On August 16, the Privy Council reviewed the original complaint and Houblon's defense. They referred the case to the Committee on Trade and Plantations, who then heard additional testimony in early September. Houblon's defensive strategy appeared to pay off: the Committee seems to have ignored the complaints of the wives and focused on the theft of the *Charles II* instead. The committee issued a formal proclamation: "Orders maybe be given that the Ship with all the Ship's Company be Stopt and Seiz'd into Safe Custody in the Plantations of where soever she shall be met with." For the first time, Henry Every and his crew were officially on the run from the law.

Houblon would continue to be plagued by legal actions into the spring of 1695, after the long-suffering remaining crew of the Spanish Expedition returned from A Coruña. Led by William Dampier, the non-mutineers appealed to the High Court of Admiralty in an attempt to extract their wages from Houblon and his fellow investors. Dampier claimed that he had been contractually promised £82 for his services and had been paid only £4. Using the most circumstantial of evidence, the expedition's sponsors successfully argued that Dampier and the other officers had resisted orders and assisted Every and the other mutineers in escaping with the *Charles II*. (This accusation is the most plausible explanation for Dampier's subsequent reticence about the entire affair—he had his eye on potential government benefactors and couldn't risk being branded a mutineer.)

The case was finally dismissed in January 1696. Dampier never recovered his back pay, but by that point, he was putting the finishing touches on his maritime memoirs, a book that would ultimately prove to be far more lucrative to him than the Spanish Expedition had promised to be.

The financial scandal of the Spanish Expedition received little attention from the London press during its initial stages. Disputes over unpaid wages were common in the shipping business; the High Court of Admiralty would usually hear more than a hundred such complaints a year. But the Spanish Expedition's legal imbroglio was also overshadowed by a far more dramatic financial crisis: the cratering stock price of the East India Company.

Boosted by the insatiable demand for calico and chintz among London's elite, the company had enjoyed a historic run of profits in the second half of the 1600s. The company transitioned to a modern-style corporation in 1657, when it began issuing general shares that gave investors a piece of the overall enterprise and just not specific voyages, and in the three decades that followed, the company completed over four hundred expeditions to India, increasingly focused on textiles. In 1670, slightly more than half its imports were cotton-based products, with pepper and other spices lagging behind for the first time. By the mid-1680s, calico and chintz constituted 86 percent of the company's trade with India. The insatiable demand for Indian cotton generated historic returns for the company's investors. Shares worth £100 in the middle of the 1660s were worth more than £500 by 1680. And the company's dividends were far more generous than those of modern corporations. For much of the second half of the century, the company returned a 20 percent dividend annually to investors, but in the boom years of the calico trade in the 1680s, the dividend reached as high as 50 percent. An investor buying £100 of shares in the company in 1657 would have received £840 in divi-

dends by 1691, on top of the increase in the value of the shares themselves. The wealth creation was not quite as dramatic as the IPOs of
the modern digital age—$100 invested in Apple's public offering in
1980 would be worth roughly $40,000 today—but in the seventeenth
century, that level of return on investment was unprecedented.

Starting in the late 1680s, however, a series of political and financial scandals began to undermine the company's economic
prospects. The company's pioneering shareholder structure was
accompanied by other, shadier innovations that would also become
commonplace in modern-day financial markets. Josiah Child, the
company's aggressive governor (the equivalent of a modern CEO),
developed a keen skill at manipulating the market with selectively
released information—with varying shades of truth and falsity—
regarding the company's fortunes in India. Years later, in his *Anatomy
of Exchange-Alley*, Daniel Defoe would describe Child's influence
over the investor class:

> The East India Stock was the main point, every Man's
> Eye, when he came to Market, was upon the Brokers, who
> acted for Sir Josiah, enquiring 'does Sir Josiah Sell or Buy?' If
> Sir Josiah had a mind to buy, the first thing he did was to
> commission his brokers to look sower, shake their heads, sug-
> gest bad news from India; and at the bottom if followed, 'I
> have a commission from Sir Josiah to sell out whatever I can,'
> and perhaps they would actually sell ten, perhaps twenty
> thousand pound. Immediately, the Exchange . . . was full of
> sellers; nobody would buy a shilling, 'till perhaps the stock
> would fall six, seven, eight, ten percent, sometimes more;
> then the cunning jobber had another set of men employed on
> purpose to buy, but with privacy and caution, all the stock
> they could lay their hands on.

Today, of course, such blatant manipulations might prompt a visit from the SEC, but in the 1680s, all the now familiar conventions of publicly traded exchanges were in the process of being invented, and the distinction between outright fraud and shrewd investing had not yet been codified into law. Part of Child's cunning came from the way he exploited the imperfect flow of information from India itself. The company's trading monopoly with India gave it an additional quasi-monopoly on news from the subcontinent, which enabled it to invent entire narratives for the purposes of manipulating the share price, with very little risk of fact-checking from a journalistic entity or a rival firm. As Defoe wrote:

> There are those who tell us, letters had been order'd by private management to be written from the East Indies with an account of the loss of ships which have been arrived there, and the arrival of ships lost; of war with the Great Mughal, when they have been in perfect tranquillity, and of peace with the Great Mughal when he has come down against the factory of Bengal with 100,000 men, just as it was thought proper to call those rumours for raising and falling of the stock and when it was for this purpose to buy cheap or sell dear.

The growing power of the company—coupled with these shady financial practices—led to increasingly strident calls for Parliament to rescind the exclusive charter it had granted the company for trade with India. Josiah Child had managed to maintain support for his enterprise by making substantial kickbacks to members of King James II's extended court, but when William of Orange ousted James in the Glorious Revolution of 1688, all those years of bribery became worthless overnight. Shortly after William's coronation, Parliament opened a series of investigations into the corrupt practices of Child and his fellow executives; a proposal emerged for a New East India Company that would compete with the original and offer shares to a wider slice of the British merchant class.

The attacks on the company in London during this period were matched by commensurate attacks in India itself. The East India Company factory at Surat—where William Hawkins had landed in 1608—had been the site of increasingly hostile disputes with representatives of Aurangzeb. Seeking more autonomy over its trading practices, the company relocated its headquarters to an archipelago of seven islands on the Konkan Coast, almost 200 miles south of Surat. The islands had once belonged to Portugal, but they had been ceded to the British as part of the dowry of the Portuguese princess Catherine of Braganza in her marriage to Charles II in 1661. Seven years later, they were leased to the East India Company, and by 1687, the company had made the archipelago its primary base in India. The Portuguese had called the islands Bombaim, which was eventually anglicized into the name they would retain until the twentieth century: Bombay.

As part of the charter granted by Charles II, the company was authorized to construct a mint in its Bombay headquarters, in part to create a more stable currency for trading than the chaotic mix of private and state-sponsored coins that the existing Indian market utilized. The company recruited a well-connected nineteen-year-old named Samuel Annesley to oversee the mint operations. Annesley was the son of a prominent dissenting minister, a member of an extended family with a rich intellectual and theological history. His father was friends with Daniel Defoe, who wrote an elegy on the occasion of his death. His nephew was John Wesley, the founder of Methodism. Annesley appears to have had more commercial than spiritual interests, however. Whatever entrepreneurial talents he might have brought to the job, Annesley arrived in Bombay to find that the mint operation had been effectively suspended. "The mint was there, and it was working after a fashion," his biographer Arnold Wright observed, "but there was no call for additional coinage. The problem indeed was how to get rid of the money that was already minted. There was no scope for its circulation outside the island, and on the island itself the needs of the bazaar were day by day undergoing greater restriction." Before long, he had been dispatched to Surat to help oversee the factory there.

In Surat, Annesley found a more established settlement than the nascent headquarters on the Bombay archipelago. The East India Company factory had grown into a collection of warehouses and residential halls, surrounded by a wall, overlooking the muddy estuary of the Tapti River. The town itself contained as many as two hundred thousand residents, and the commercial activity it attracted as the hub of Red Sea trade generated enough wealth to support numerous mansions on its finer streets, along with "marble seraglios, beautiful scented gardens and plashing fountains, and a general exuberance of costly adornment." Annesley integrated himself quickly into the broader trading community. Within a few years, according to Wright, "he had a profound knowledge of all the ramifications of the Company's business at Surat; he was intimately associated with the native trading community; and he was acquainted with all the network of intrigue which enmeshed the official life of the place." His primary complaint with his existence in Surat lay in the deep-rooted association in the minds of the local trading community—and the Mughal authorities—connecting the East India Company with the pirates of the Red Sea. "Was it not on account of the piracies," Annesley wrote in one letter back to the Bombay headquarters, "we should live here in as great, or greater honour, credit and respect than ever."

The association with the Red Sea pirates also bedeviled the

company representatives—known as "factors"—back in Bombay, though they had other threats to contend with as well. By shifting its operations to Bombay Castle, the company had hoped to enjoy far greater freedom from the whims of Aurangzeb. ("Though our business is only trade and security, not conquest," the company directors declared, "yet we dare not trade boldly or leave great stocks where we have not the security of a fort.") But their locus in a subtropical marshland made them vulnerable to a different kind of local hazard: disease. One visitor in 1690 found the British merchants overwhelmed with "pestilential vapours that seized their vitals and speeded their hasty passage to the other world." By some reports, more than half the settlement died during the epidemics that swept across the archipelago. Aurangzeb, meanwhile, continued to challenge the company's trading authority. He briefly imprisoned the company representatives in Surat, and laid siege to Bombay Castle, ransacking its warehouses situated outside the castle's walls. Negotiations ultimately secured a truce, with the company compelled to pay a fine of 150,000 rupees and to promise to "behave themselves for the future no more in such a shameful manner."

The fragile détente with Aurangzeb was immediately followed by more hostilities back home. In 1693, the House of Commons voted to dissolve the original corporation and start anew with a more open approach to Indian trade. But just as the company seemed to be on death's door, William reversed course and renewed the charter while Parliament was on leave. Enraged by the betrayal, Parliament quickly passed a resolution that declared "all subjects of England should have equal right to trade to the East-Indies." In May 1695, as the Spanish Expedition Shipping case was winding to its close, Parliament began another investigation into corruption at the highest levels of the East India Company. "Even by the lax standards of the day," Nick Robins writes in his history of the East India

Company, "politicians were genuinely shocked by what they found. A team of MPs pored through the company's accounts and uncovered a complex web of bribes, all emanating from the Governor Sir Thomas Cooke, Child's son-in-law. In the six years since the Revolution, £107,013 had been paid out for 'the special Service of the Company,' including a massive £80,468 in 1693 to win a new charter."

Understandably, the constant turmoil and uncertainty had a catastrophic effect on the company's share price, which dropped 35 percent over the course of 1695. It would decline almost as much the following year.

All these developments—the corruption investigations, the still fragile relationship with Aurangzeb, the threat of losing the company's exclusive charter, the cratering stock valuation—weighed heavily on the mind of Samuel Annesley, now thirty-seven years old and recently appointed president of the Surat factory. The news from back home rolled in with a lag of a month or two, but by midsummer 1695, he would have been able to detect the broad strokes of the story: the hordes were coming after the East India Company back in London.

They would be coming for Surat soon enough.

13

WEST WIND DRIFT

The Atlantic Ocean, west of Africa

Late 1694

T he *Fancy* had departed from A Coruña with ample supplies: 150 tubs of bread, 100 muskets, and over 100 barrels of gunpowder. Every's most pressing need was not food or armaments, but rather men. The eighty pirates aboard the ship would not give Every enough manpower to exploit the *Fancy*'s full potential in an exchange with another vessel. Each of the great guns on deck required at minimum six men to man them in the heat of battle; with forty-six guns on board, Every knew he needed at least three times his current crew to fire a full broadside. More would be needed to fire the muskets, man the sails, and storm enemy ships if the *Fancy* were able to overpower them.

After three weeks sailing south-southwest past Portugal and the Strait of Gibraltar, then along modern-day Morocco and the western edge of the Sahara, Every and his men made their first stop in Cape Verde, the archipelago of ten volcanic islands situated 350 miles off the coast of Africa. There they raided three English vessels for provisions and briefly detained the Portuguese governor in order to ensure that they could plunder with impunity. They took on fish, beef, salt, and "other necessaries." Preeminent among those necessaries were nine sailors from the British ships who opted to sign on with Captain Every and his crew.

While the stopover at Cape Verde marked the second criminal act of Every's life as a pirate, it also displayed a quality that would be increasingly evident in his actions: a deliberate attempt—if arguably a futile one—to retain some semblance of ethics and legitimacy, particularly with regard to English property. Just as Every had allowed dissenters—including Captain Gibson—to disembark the morning after the mutiny, and tossed a bucket to them when their longboat began taking on water, Every had his quartermaster, Joseph Dawson, write up bills for everything they had pilfered from the English ships, which he left with the victims, presumably with some promise to recompense them for their loss.

For the crew of the *Fancy*, the Cape Verde sojourn represented an important crossroads. The ship was perfectly situated to ride the trade winds due west across the Atlantic, following the path that Atlantic hurricanes take every late summer and fall. An "extraordinary sailor" like the *Fancy* could be in Barbados within a matter of weeks. Many of the men would have made that exact journey multiple times already in their careers at sea. But Every had a far more ambitious—and challenging—expedition in mind: sailing due south and rounding the Cape of Good Hope, then venturing up the eastern coast of Africa to Madagascar. That journey was far riskier. With the provisions they had "borrowed" from the English ships, the *Fancy* had plenty of supplies to make her way to the West Indies. A voyage to Madagascar would take months, and require multiple stops for provisions. And rounding the cape itself—with its rogue waves, strong currents, and treacherous coastline—posed great danger even for experienced sailors.

Why was Every so determined to follow such a perilous itinerary? There was plenty of gold to plunder in the West Indies, after all. But Madagascar offered something far more tantalizing than Spanish galleons traversing the Caribbean. Famously friendly to pirates, the island served as a launching pad for voyages into the Indian Ocean, where buccaneers could prey on treasure ships making pilgrimages to Mecca. The wealth of the Mughal Empire had reached such legendary proportions that it had captured the imagination of a new generation of pirates, soon to be called the "Red Sea Men" for their frequent raids at the mouth of the Red Sea. Trapped in bureaucratic limbo in the A Coruña harbor, or anchored off the arid islands west of Africa, Every could still hear the siren song calling him from the other side of the world: the treasure of a five-centuries-old dynasty, floating in the quiet waters of the Red Sea. The sunken galleons and trade ships of the West Indies were nothing compared to the riches of Aurangzeb.

If Every's men required persuading to make the journey around the cape, the recent exploits of the American pirate Thomas Tew would have provided ample motivation. At the time, Tew was arguably the most famous pirate in the world, though his celebrity was nothing compared to what Every was about to achieve. Raised in Rhode Island, Tew had attracted a small retinue of investors in Bermuda, who helped him secure an eight-gun sloop called the *Amity*. Armed with a letter of marque from the governor of Bermuda—the very same man Henry Every had worked for as a slave trader—Tew and his men crossed the Atlantic and rounded the Cape of Good Hope, eventually making their way to the Red Sea, where they encountered an Indian trading vessel, which immediately surrendered to them. Tew and his men walked away with more than £100,000 worth of silver, gold, spices, and fabric. With only forty-five pirates aboard the *Amity*, the profit sharing among the crew was bountiful:

most of the men on board took home roughly £2,000 after the prizes had been fully allocated by the quartermaster. Recall the terms that James Houblon had offered the experienced crew of Spanish Expedition Shipping: £82 for the entire voyage. A midshipman on the *Amity* had earned fifty times that in a six-month voyage. You could make a decent living working as a senior officer aboard a commercial vessel like the *Charles II* in its original incarnation, salvaging treasure from sunken ships in the Sargasso Sea. But you could make a fortune as a Red Sea pirate.

There is some speculation that Henry Every might have known Thomas Tew personally, given their mutual connections to the governor of Bermuda. Every might have invoked that association as part of his case for a Red Sea expedition: *When I last saw Thomas Tew,* he might have told his men, *he was scrounging around Bermuda, trying to put a venture together. Now he is rich beyond imagination. That could be our story, too.*

Whatever case Every made to his men, by the time the *Fancy* left the Cape Verde islands, they were committed to the Red Sea plan. They sailed south along the edge of Guinea, anchoring just offshore, close to coastal settlements that were known to engage in commerce with Europeans. Every ordered the crew to "put out English colors to make the natives come aboard to trade," Philip Middleton would later recall. They may have lured the Guinean villagers onto the ship by leaving small goods on the beach to attract the interest of the community. In an oral history of the slave trade, the descendent of a West African captured during this period described the deceit the Europeans used to apprehend her ancestor:

> One day a big ship stopped off the shore and the natives hid in the brush along the beach. Grandmother was there. The ship men sent a little boat to the shore and scattered

bright things and trinkets on the beach. The natives were curious. Grandmother said everybody made a rush for them things soon as the boat left. The trinkets was fewer than the peoples. Next day the white folks scatter some more. There was another scramble. The natives was feeling less scared, and the next day some of them walked up the gangplank to get things off the plank and off the deck.

The Guineans had more reason to trust the strange new English ship in their waters than we might now expect. At the end of the seventeenth century, the slave trade was still dominated by the Spanish and the Portuguese; the Royal African Company had only just recently shifted its focus from gold to enslaved humans. But the *Fancy*'s posture as a merchant vessel turned out to be a trap. Boarding the ship with the hopes of exchanging goods with the Europeans, the Guineans suddenly found themselves imprisoned. According to Middleton, "when they came aboard, [the crew] surprised them, and took their gold from them, and tied them with chains, and put them into the Hold." On the shore, the remaining members of the Guinean village cried in horror as the British ship raised anchor, their loved ones slowly disappearing over the horizon in the English vessel, never to be seen again.

What kind of life did those captured Africans experience on board the *Fancy*? The historical evidence for this is murky. From Middleton's description, it appears that—initially at least—the Africans were treated as prisoners, potentially as goods to be bartered at another port of trade. We know that seven of the captured Africans would later be sold into slavery, but it is unclear what happened to the others—assuming there were others. It was not unheard of for pirates during this period to free captured slaves, and grant them equal rights as members of the pirate commune. Recent scholars have argued that the pirate crews that terrorized the West Indies were surprisingly multiracial in their composition, with Africans constituting more than 20 percent of the onboard population. As the historian David Olusoga notes in his book *Black and British*, Francis Drake's 1577 expedition around the world was "achieved with a crew that was what we would today call inter-racial," despite the fact that Drake himself was also a slave trader. "In a way we find difficult to relate to," Olusoga writes, "[Drake] was capable of enslaving black people while seeing other black men as his comrades-in-arms." In later pirate expeditions, several freed slaves rose to high-ranking positions, most notably Black Caesar, who, according to legend, was a former African chieftain who served as a lieutenant to Blackbeard on the *Queen Anne's Revenge*. As working members of the crew, the African pirates would have participated in all the proto-democratic conventions of the pirate collective, which made the pirate ships of the golden age the first Western institution to extend suffrage to people of color.

Nothing, however, in the historical record—and in Henry Every's previous career as an interloper in the slave trade—suggests that the captured Africans on board the *Fancy* were granted similar privileges. On a ship that already posed significant challenges to even the high-ranking officers, they most likely suffered through a brutal existence: manacled belowdecks for days on end, interrupted by stretches of forced labor, all the while wondering what these haggard Europeans had in store for them. Like most pirate ships of the age, the *Fancy* was a floating commune, a seedbed for radical ideas about wealth sharing and democratic governance. She was also, inarguably, a slave ship.

The *Fancy* continued southeast to Fernando Po, an island south of modern-day Nigeria that is now part of Equatorial Guinea. There Every instructed the men to make significant alterations to the ship, removing most of the upper decks, including the captain's quarters in the sterncastle. These were customary alterations for pirates to make after seizing a ship. Creating a flush upper deck significantly reduced wind resistance at sea, making a fast ship like the *Fancy* even faster and easier to maneuver while engaged with potential enemy ships. Eliminating structures on the deck would also make it easier for them to pump water if the ship encountered high waves in the turbulent seas around the Cape of Good Hope. The alterations had political undertones as well: in order to maximize both agility in the water and manpower on board, most pirate captains disavowed their exclusive quarters and slept with the rest of the crew belowdecks. The egalitarian ethos of the pirate community extended to the architecture of the ship itself.

At Fernando Po, the crew also spent weeks sweating through the tedious work of careening the hull of the *Fancy*. When we think of the existential threats eighteenth-century pirates faced, our mind naturally conjures up enemies who want to sink their ships with cannon fire. But as Every and their men had sailed the warm Atlantic waters off of Africa, their one truly unavoidable nemesis was already clinging to their ship, lurking below the surface: shipworms feasting on the *Fancy*'s timber.

Despite their name and appearance, shipworms are, in actuality, mollusks; they are effectively clams disguised as worms. Shipworms burrow into wood that has been submerged in water, releasing bacteria that digest the wood's cellulose. Left unattended, shipworms could destroy the hull of a ship like the *Fancy* in four or five months, as Henry David Thoreau would later describe in his poem "Through All the Fates":

> Far from New England's blustering shore,
> New England's worm her hulk shall bore,
> And sink her in the Indian seas . . .

The renegade nature of pirate life meant that a ship like the *Fancy* almost never enjoyed access to a dry dock where her hull could be easily repaired. Careening was the only option: deliberately grounding a ship on a beach at high tide, exposing one side of the hull so that the damaged inflicted by shipworms—and by other quiet nemeses like barnacles and rot—could be undone. In tropical waters, seaweed clinging to the hull could increase drag as well. Every captain in command of a wooden ship—which, in the late seventeenth century, meant every captain, full stop—had a countdown clock running in the back of his mind, recording the weeks or months that had passed since his last opportunity to careen his vessel. If you found yourself on a wooden ship trapped in the middle of a vast ocean with no wind, you could be killed by a mollusk as readily as you could die of thirst.

By early fall of 1694, with the *Fancy* now snug and streamlined for speed and agility, her hull newly repaired, Every and his men could attempt the journey around the Cape of Good Hope. Shortly after leaving Fernando Po, they engaged in a brief skirmish with two Danish privateers, the first test of the *Fancy*'s armaments. The Danish quickly surrendered, and the pirates took on "forty pounds of gold dust, chests of fabrics, small arms and fifty large casks of brandy." Seventeen of the Danish privateers found Every and his "stout frigate" compelling enough to join the crew of the *Fancy*, which now numbered almost a hundred.

To round the cape toward Madagascar, Every needed initially to sail in the opposite direction, away from the western edge of Africa, almost all the way across the southern Atlantic to Brazil. There he could catch a ride on one of the planet's most powerful conveyer belts: the Antarctic Circumpolar Current, also known as the West Wind Drift. Utilized by Vasco da Gama in his pioneering voyage around the cape, the West Wind Drift is a vast but leisurely flow of cold water, more than twice the volume of the Gulf Stream, that propels ships from west to east, in southern latitudes safely below the menacing rocks of the cape. The collision between the colder Antarctic waters and the warmer waters of the South Atlantic creates extensive nutrient upwelling along the route, supporting a rich ecosystem of sea life. The crew of the *Fancy* enjoyed the sight of whales, seals, penguins, and albatrosses as they made their way east toward Madagascar. By riding the West Wind Drift in the southern hemisphere's summer months, they had little risk of colliding with icebergs.

Sometime in the first months of the new year—1695—the lookout aboard the *Fancy* spotted the distinctive clawlike sandspit that extends along the western edge of Saint-Augustin Bay, on the southern coast of Madagascar. The *Fancy* had made it to safe harbor in the Indian Ocean, on an island widely recognized as a haven for pirates. After a voyage of roughly five thousand miles, it was time to enjoy a few months on land, and prepare for the next act.

14

THE *GANJ-I-SAWAI*

Surat, India

May 1695

As the crew of the *Fancy* careened their vessel on the shores of Madagascar, across the Indian Ocean, in the harbor at Surat, another ship was taking on provisions for a different kind of voyage. This ship was a *ghanjah dhow*, or wooden trading vessel, owned by the Grand Mughal Aurangzeb himself. A visitor surveying the Surat harbor skyline from the other side of the Tapti River would have been able to make out the ship easily from a distance, a giant looming over the galleys and East Indiamen anchored along the banks of the river. At 1,500 tons, with enough room on board to accommodate over a thousand passengers, she was almost certainly one of the largest ships in the world at the time. Aurangzeb had given her the name *Ganj-i-Sawai*, Persian for "exceeding treasure." In the news reports and court trials and popular lore that circulated through the English-speaking world, the ship's name would be anglicized into a simpler form: the *Gunsway*.

The *Gunsway* was based out of Surat, along with four smaller vessels belonging to the Grand Mughal that often sailed alongside her.

Aurangzeb had commissioned the ships for an explicit purpose: to transport dignitaries—some of them members of his immediate family—to Mecca for the hajj, the annual pilgrimage of Muslims to the holy lands at the base of the Asir Mountains, east of the Red Sea. Along the way, the *Gunsway* and her escort would stop over at the trading port of Mocha, near the mouth of the Red Sea in modern-day Yemen. With the newfound craze for coffee raging across the capitals of Europe, Mocha enjoyed a flourishing economy as one of the central nodes in the international coffee trade. (Modern consumers savoring their Mocha Frappuccinos at Starbucks pay a distant tribute to the city with each order.) The coffee beans attracted traders specializing in other goods as well, giving Aurangzeb an additional commercial incentive to send the *Gunsway* on the pilgrimage—a fitting mix of business and piety for a religion that had been founded by a trader a thousand years before.

The manifest for the *Gunsway* would have been an formidable document. The ship's hull was loaded with calico textiles, fine porcelain, ivory ornaments, and other valuables. Along with the food required to keep pilgrims and crew alive, the *Gunsway* carried barrels of spices to trade in Mocha, predominantly peppercorns. The contemporary mind might find something amusing in the idea of a treasure ship weighted down with a condiment that is now so cheap that we give it away for free at restaurants, but in the seventeenth century, pepper was still one of the most highly sought-after luxury goods in the world. Its price had declined from its peak in the Middle Ages, when peppercorns were often worth significantly more than their weight in gold, but even with the decline the pepper barrels could be traded for a fortune at Mocha. Eighty cannons lined the main deck, manned by more than four hundred soldiers, protecting both the treasure and the eight hundred pilgrims on board.

Making the pilgrimage to Mecca during the hajj constitutes one of the five pillars of the Islamic faith. (The others are professing faith in the one God and Muhammad as his prophet, prayer performed five times a day, charitable giving, and fasting during Ramadan.) Observant Muslims must participate in the hajj at least once in their lifetime, during the final month of the Islamic calendar. Today, Mecca is a Saudi city with roughly two million inhabitants that, amazingly, triples in size during the hajj. The influx of pilgrims each year is the single largest annual migration of human beings on Earth. (Far more people travel annually during Chinese New Year, but they are distributed in rural regions across China, not converging on a single destination as they do in the hajj.) Each year, the Saudis erect an immense pop-up city outside Mecca consisting of 160,000 air-conditioned fiberglass tents, each housing fifty pilgrims, a desert settlement that makes the temporary housing of Burning Man look like a shantytown.

Because the Islamic calendar follows a lunar cycle, each Islamic year is approximately eleven days shorter than a year following the Gregorian calendar, which means that the actual timing of the hajj shifts backward from year to year. Measured by a Western calendar, a hajj that begins January 1 would be followed the next year by one that commences on December 20. In 1695, the last month of the Islamic calendar corresponded to July on the Gregorian calendar, which meant that the voyage from Surat to Mecca—roughly the same distance as sailing from Istanbul to Gibraltar—would need to begin in late spring to give the traders on board sufficient time to do business in Mocha and other port cities along the way.

The tradition of the hajj dates back to Muhammad's conquering of Mecca in 629 CE, during which he destroyed pagan icons in an ancient granite temple known as the Kaaba, declaring "Truth has come and Falsehood has Vanished." After reconsecrating the building as a shrine to Allah, Muhammad then led a pilgrimage from

Medina to Mecca in 632, where he delivered his farewell sermon. But the religious significance of the site predates Islam. According to the Quran, the Old Testament figure of Abraham (also considered a prophet in the Islamic tradition) is commanded by God to take his child Ismail (Ishmael in the Old Testament) and his wife Hagar out to a bleak patch of desert that marks the site of modern-day Mecca and leave them there to die of thirst, as a test of his faith. After days of intense suffering, a well miraculously appears in the arid landscape, saving mother and child at the last minute.

If you are a religious person—regardless of your faith—the long chain of influence generated by that experience in the desert five thousand years ago makes a certain kind of sense, no matter what God you happened to worship. When a supreme being has direct contact with a mortal, it makes sense that ripples would continue to spiral out from that encounter fifty generations later. But if you *don't* believe in supreme beings, the chain of influence is baffling. Someone has a dream in the desert of a divine presence who commands him to murder his wife and child, and seven thousand years later, *six million people* travel to the foothills of a desert mountain range once a year to visit the place where it all happened. There are very few echoes in the cathedral of history that have reverberated for so long with such faint origins.

The emergence of pilgrimages as a mass ritual—Muslim or otherwise—marked a watershed in the lived experiences of ordinary people. In an age before tourism, pilgrimages introduced long-distance travel to millions of human beings who would otherwise have spent their entire lives on a much smaller patch of land. By 1695, the hajj had grown into one of the planet's great melting pots, creating a shared space where North Africans, Arabs, Europeans, and Indians could converge for one month out of the year. They were there to pray, but they were also there for something else, something we would now call *the scene*. Some of the richest people in the world took months out of their calendar to journey up or down the Red Sea to pray in front of the Kaaba; many of them planned their entire year around the journey, the way so many of us today plan our calendars around summer vacation. The grandeur of the *Gunsway* was not simply an attempt to create a luxurious cruising vessel for the Grand Mughal's inner circle. It was also a statement to the world, like a billion-dollar luxury yacht that pulls into harbor for a revival meeting, a way of broadcasting the scale of the Universe Conqueror's fortune to other pilgrims who would never visit the sublime architecture of Agra or see the Peacock Throne in Delhi.

Of course, transporting all that wealth to a city thousands of miles away in a foreign nation made the ship supremely vulnerable. The eighty guns and four hundred soldiers aboard the *Gunsway* were there for a reason. But the risk was compounded by the geography of the region. The Red Sea empties out into the Gulf of Aden through a narrow strait just twenty miles wide called the Bab-el-Mandeb. Because the Suez Canal was still centuries away, in 1695 any vessel making the pilgrimage to Mecca—or trading with the port cities of the Red Sea—needed to pass through the strait of Bab-el-Mandeb and the Gulf of Aden before entering the wider expanse of the Arabian Sea. In modern times, the immense wealth that passes through those straits takes the form of oil, loaded up in the refineries that line the Red Sea. In 1695 the wealth had a different manifestation: jewels, spices, gold, cotton. But then, as now, the bottleneck of the Bab-el-Mandeb made the region uniquely suited for piracy. It is no accident that the Somali pirates, the most notorious of the twenty-first century, operate out of the exact same waters.

Everything that made the strait of Bab-el-Mandeb and the Gulf of Aden such an advantageous route for traders also made it a hunting ground.

15

THE *AMITY* RETURNS

The Gulf of Aden
Spring 1695

The charms of Madagascar were not generally apparent to the first generations of Europeans to visit the island. One observer described the place as "Swarms of Locusts on the Land, and Crocodiles or Alligators in their Rivers." In 1641 an Englishman named Walter Hamond became so enamored with the island and its native Malagasy people (he wrote a pamphlet calling them "happiest people alive") that he led a group of English Puritans to build a community at Saint-Augustin Bay—the Indian Ocean version of the *Mayflower* Puritans who had helped settle Massachusetts two decades earlier. Hamond inaugurated what would prove to be an enduring literary tradition of Europeans spinning elaborate fantasies of an island utopia off the eastern shore of Africa, a tradition that Every would play a central role in as well. In one of his missives, Hamond called Madagascar the "richest and most fruitful island in the world." It is unclear how many of his fellow settlers shared his opinion. The colony had disintegrated by 1646.

Other Europeans tried to get a foothold in the years that

followed. The French established Fort Dauphin to the east of Hamond's settlement. The Portuguese extracted slave labor from the native populations where they could. But the island retained its autonomy, along with a certain reputation for lawlessness. By the time Henry Every arrived there in the early months of 1695, Madagascar was a pirate's den.

At Saint-Augustin Bay, and in other secluded harbors to the north, the crew of the *Fancy* enjoyed a productive idyll as they prepared for their Red Sea assault. It is unclear whether Every was aware of the exact timing of the hajj that year, or whether he simply knew that the western monsoon winds of late summer were likely to generate a significant amount of shipping traffic in the Gulf of Aden in August. Either way, he seems to have recognized that the most propitious time for an attack would not arrive until summer. Biding their time, the crew careened the *Fancy* again. They savored the Danish brandy they had pilfered back on Cape Verde. They traded a few guns and some gunpowder to the Malagasy for a hundred head of cattle, and spent most of March feasting on roast beef. By late spring, they had sailed for the Comoro Islands, where they enticed another forty men from a French ship to join their number. (After plundering the French ship for rice, they sunk it in the harbor, which may have made their case slightly more persuasive.) The pirates also bartered for hogs and vegetables, before escaping to sea after three East India Company ships appeared on the horizon.

With more than 150 men under his command, and the summer months rapidly approaching, Every decided it was at last time to execute the plan he had been mulling over since those long days and nights in the A Coruña harbor. The *Fancy* sailed along the coast of modern-day Somalia, headed toward the Gulf of Aden. Stopping over in a town the pirates called Meat—in actuality, Maydh—their efforts at trading were rebuffed by the local Muslim community.

"The people would not trade with us," the ship's coxswain John Dann would later say of the town, "and we burnt it." According to some accounts, the pirates went so far as to plant gunpowder beneath the local mosque, demolishing it as an act of revenge.

That demolition raises an interesting question: To what extent were Every and his men animated by the fact that they had set their sights on a target that was specifically Muslim—those "Moor" ships making their annual pilgrimage? Was their mercenary desire for treasure enhanced (or legitimized) by the idea they were also going to be waging war on the infidels? Certainly they would have described themselves as "anti-Muslim," if you had asked them. But was that a core faith, or just a convenient one?

It is difficult to say from such a distance. On the one hand, the crimes Every's men would later commit on board one of those Moorish ships were horrendous ones, ones they might well have refrained from had they captured a ship populated by Christians. But the fact that they were targeting Muslim vessels in the first place had an obvious financial justification. To paraphrase the classic Willie Sutton line about bank robbery, the Muslim ships were where the money was.

But the mosque at Maydh was different. There was nothing to be gained by destroying it. Yes, as pirates, you might resort to violence (or the threat of it) to coerce a town that refused to trade with you to hand over whatever you had hoped to barter for. But going out of your way to plant explosives beneath a mosque suggests a deeper level of contempt. It seems likely that at least a few of the key men on board the *Fancy*—if not Every himself—harbored actively anti-Islamic views.

Entering the Gulf of Aden, it quickly became apparent that Every was not alone in his scheme to prey on the Red Sea pilgrims. First, they encountered two American privateering ships, the *Dolphin* and the *Portsmouth Adventure*, with a combined crew of 120 men. Together, they sailed to Perim, a crab-shaped volcanic island in the strait of Bab-el-Mandeb. "We lay there one night," Philip Middleton would later recall, "and then three more came. One commanded by Thomas Wake fitted out from Boston in New England; another, the *Pearl* Brigantine, William Mues Commander, fitted out of Rhode Island; the third was the *Amity* sloop, fitted out at New York. They had about six guns each. Two of them had 50 men on board and the Brigantine between 30 and 40." The *Amity* was no stranger to these parts. At her helm was the legendary Thomas Tew, the pirate whose successful Red Sea heist two years before had inspired Every's original scheme.

The convergence of all these pirates—traveling thousands of miles independently to arrive on the same tiny island in the mouth of the Red Sea—tells us something about just how irresistible the siren song of the Grand Mughal's wealth was in 1695. All told, the six ships held 440 men. In the summer of 1695, they probably represented a significant fraction of all working pirates on the planet. During the golden age of piracy in the early 1700s—when a generation of buccaneers inspired by Henry Every wreaked havoc in the Caribbean—official estimates at the time put the total global pirate population at roughly two thousand. Assuming that number is higher than the global pirate population in 1695, before Every's mythological story drove the next generation to sea, the pirates clustered together in the strait of Bab-el-Mandeb that June might well have represented fully half of all the pirates on the seven seas at that moment in history.

No doubt Every experienced mixed feelings watching these other vessels sail into his hunting grounds. On the one hand, they would all be vying for the same treasure, and after an entire year as captain of the *Fancy*—with largely unchallenged control over his own destiny—his future actions might well be mediated by the leadership on board the other ships. But on the positive side, Every had been struggling from the beginning to build up enough manpower to take on the well-fortified ships of the Indian fleet. If the six ships anchored in the Perim harbor could agree to work together, they might well have enough force to challenge Aurangzeb's mightiest vessels.

According to Middleton's testimony, the six captains convened after consulting with their crews and set out the terms for their alliance. Articles of agreement were almost certainly drawn up. The fact that they would consolidate forces at this critical moment was not, in itself, surprising. If they managed to overtake one of the Indian treasure ships, there would be plenty of loot to go around. What *was* surprising about the alliance between the six ships was their choice of a leader. Thomas Tew was, on paper, the obvious candidate. Every had done most of his trade on the western coast of Africa, and was by all accounts a newcomer to piracy as a full-time profession. Tew had just pulled off a heist of historical proportions *in the very waters* they were currently sailing. But something in the chemistry of those two men, and the crews they commanded, led to a different outcome. "They all joined in partnership," Philip Middleton reported, "agreeing Captain Every should be the Commander."

For twelve months, Henry Every had kept his men alive and eluded capture with little more than a fast boat and a cunning plan at his disposal.

Now he had an armada.

16

SHE FEARS NOT WHO FOLLOWS HER

Bombay
May 1695

John Gayer had barely assumed the position of governor of the East India Company, overseeing all of the company's operations in the subcontinent, when news of a pirate named Henry Every first came across his desk. The son of a merchant, and nephew of London mayor Sir John Gayer, he had grown up not far from Every's hometown in Devonshire. Like Every, he took to the sea as a young man, and soon became a ship's captain for the East India Company. In the early 1690s, he had been dispatched to the subcontinent with 120 English soldiers with instructions to stabilize the "depleted" Bombay port and factory; just two years later, he was overseeing the entirety of the company's Indian business.

In May of the following year, as Every's men were lighting those explosives under the mosque at Maydh, a report arrived courtesy of the three East Indiamen that had chased Every off the Comoro Islands. It was a factual account of their interactions with Every, but it was also a prediction: Every would be a problem for the East India

Company. Much of the document echoed a refrain that appears throughout Every's run as captain of the *Fancy*: *This pirate is commanding an astonishingly fast boat.*

> Your Honor's ships going into that island gave him chase, but he was too nimble for them by much, having taken down a great deale of his upper works and made her exceeding snugg, which advantage being added to her well sailing before, causes her to sail so hard now, that she fears not who follows her. This ship will undoubtedly [go] into the Red Sea, which will procure infinite clamours at Surat.

The news must have seemed ominous to Gayer, given the East India Company's troubles back in London and its tortured relations with Aurangzeb. Now he had to worry about a band of pirates bound for the Red Sea in a ship that sails "so hard now, that she fears not who follows her"?

But however "snugg" Every's ship might have been, the evidence suggests he was not entirely without fear. Along with their report, the merchantmen delivered a letter they had recovered at Johanna Island, a letter penned by Henry Every himself, and left behind as an open declaration to "all English Commanders." The exact text of that letter—down to the idiosyncratic, semiliterate spelling and punctuation—survives to this day:

> *To all English Commanders lett this Satisfye that I was Riding here att this Instant in ye Ship fancy man of Warr formerly the Charles of ye Spanish Expedition who departed from Croniae ye 7th of May. 94: Being and am now in A Ship of 46 guns 150 Men & bound to Seek our fortunes I have Never as Yett Wronged any English or*

Dutch nor never Intend while I am Commander. Wherefore
as I Commonly Speake wth all Ships I Desire who ever
Comes to ye perusal of this to take this Signall that if you
or aney whome you may informe are desirous to know wt
wee are att a Distance then make your Antient Vp in a Ball
or Bundle and hoyst him att ye Mizon Peek ye Mizon
Being furled I shall answere wth ye same & Never Molest
you: for my Men are hungry Stout and Resolute: & should
they Exceed my Desire I cannott help my selfe.

    as Yett

    An Englishman's friend,

    At Johanna February 28th, 1694/5

Henry Every

Here is 160 od french Armed men now att Mohilla
who waits for Opportunity of getting aney ship,
take Care of your Selves.

What is the meaning of this declaration? On the most literal level, it conveys a code, a kind of nautical secret handshake: Bundle your flag ("Antient" here refers to the ship's ensign) into some kind of ball, and raise it to the top of your mizzen mast, and I will leave you alone. But the declaration was also a lie. He *had* ransacked British ships, albeit with more courtesy than he had shown the residents of Maydh. The historian Joel Baer interprets the letter as a "shrewd tactic to avoid conflict with the only force in the Indian Ocean capable of effectively opposing the *Fancy*, the heavily armed ships of the East India Company."

What light does this lone document shed on Every the man? From the mutiny itself to the collective decision to appoint Every—and not Thomas Tew—as admiral of the pirate armada, it seems clear that Every possessed extraordinary charisma, a "born leader of men" as one historical account has it. But what order of villain was he? As captain of the *Fancy*, he had overseen acts of genuine barbarity, from the slaves captured in Guinea to the demolished mosque at Maydh. But he had also tried to adhere to some improvised code of honor in his dealings with British citizens. The Johanna letter captures that tension exquisitely: a man who has embraced the pirate's existence outside the boundaries of nation-states and their laws, who is at the same time trying to preserve some legitimacy (and protection) as an English citizen. The letter suggests a man trying to invent on the fly a new set of codes of conduct, not a man who has renounced codes altogether. He had led a mutiny aboard an English vessel and absconded with the property of English citizens; that was undeniable. But the *Charles II* had belonged to a private venture; it wasn't as though he had stolen a Royal Navy ship. And James Houblon had reneged on the terms of their contract, leaving the men languishing in A Coruña, unpaid, for months. Every could well have convinced himself that the mutiny was within his rights, once the Spanish Expedition backers had failed to deliver what they had promised. But perhaps that attitude was purely opportunistic. Did he truly think in his own mind he was engaged in legitimate actions as a British subject, and thus should be unmolested by the East Indiamen or any other representative of the Crown? Or was that all just an act to keep the authorities at bay long enough for him to carry out his plan? Certainly he was shrewd. Certainly he was a thief. Whether there was honor in the thief—that is harder to detect at such a distance. The signal is too faint.

A meaningful slice of that signal, however, persists in the Johanna letter, particularly in its enigmatic last line: *my Men are hungry Stout and Resolute: & should they Exceed my Desire I cannott help my selfe*. The words are unmistakably intended as a threat. *My men are hungrier than yours*, they implied; *don't challenge us.* But it is not hard to read another layer into Every's words: his men may be capable of acts that he himself will not be able to prevent, even if they go against his wishes as commander. He was captain of a floating democracy, after all. Whatever power he possessed originated in the men who had given it to him. Perhaps he had already detected a capacity for violence among his crew that had disturbed him. Perhaps he recognized that their "hunger" posed its own distinct threat to his plan, that all his careful positioning could be undone by the frenzy of an out-of-control crew.

Whatever Every's intended meaning, the lines were prophetic. The hunger of his men would lead them to the most brutal and sadistic extremes of human violence. That much we know for certain. Whether they "exceeded the desire" of Henry Every in committing those acts is a harder question to answer.

17

THE PRINCESS

Mecca
June 1695

The treasure aboard the *Gunsway* was not the only thing that made the Mughal ship unusual for its time. A quick survey of the passenger manifest would have revealed another startling fact: there were dozens of women on board, many of them members of Aurangzeb's court.

The maritime world circa 1695 was overwhelmingly a world of men. Merchant ships, warships, privateers—most of them would have been entirely devoid of women. Migrant ships—like the *Mayflower*—would occasionally carry women and girls to their new homes across the sea, but a ship containing a large number of aristocratic women was almost unheard of. Not all the females aboard the ship were of noble birth; the captain had purchased a collection of Turkish concubines during the voyage and was importing them back to India, an operation that we would now consider sex trafficking. But most of the women on board the *Gunsway* were there as religious pilgrims, fulfilling their duties as good Muslims observing the hajj.

One of those pilgrims, making what would have likely been her first voyage to Mecca, was rumored to be the granddaughter of Aurangzeb himself.

The identity of this Mughal princess is shrouded in mystery. According to the official records, Aurangzeb had ten children by multiple wives, but none of those children had daughters who appear to fit the profile of the Mughal princess aboard the *Ganj-i-Sawai*. Most likely she was a member of Aurangzeb's extended family, not a direct descendant. But the mystery behind her identity itself reveals a larger point about the way histories of this period are conventionally told. The presence of so many women making their pilgrimage to Mecca is a fulcrum in the story of the *Ganj-i-Sawai*; their fate transformed what might have been a minor contretemps into a global crisis. And yet, while countless pages have analyzed the daring and savagery of Every and his men, the entrepreneurial ambition of the East India Company agents, and the wrath of Aurangzeb, the women on board the ship have only the briefest moment in the spotlight. The only identity they are granted is to be victims of a heinous crime. They have no names, no histories.

Despite that blank spot in the historical record, we can reconstruct something of what the experience of a woman in the Mughal court would have been like, and perhaps even imagine what might have been going through the mind of that young princess as she made her way back from Mecca. At the highest echelons of court society, women could play a role in political and cultural affairs, possess their own property, and even dabble in commerce. During the more progressive regimes of Akbar and Jahangir, wealthy noblewomen engaged in trade, and in some cases owned their own vessels. (In 1613 Portuguese traders seized a Mughal vessel called the *Rahimi*, at the time the largest ship in the Indian fleet. The *Rahimi* belonged to the mother of the emperor Jahangir, creating an international dispute that anticipated the crisis that would erupt around the *Gunsway* seven decades later.) Princesses could be patrons of the arts and architecture; a number of public gardens in modern-day India were originally championed by women in the Mughal court.

Some women were even active participants in the political arena. Aurangzeb's sisters had played a major role in the palace intrigue that surrounded his violent ascent to the throne. Some consider Aurangzeb's sister Roshanara Begum to be the mastermind behind his attacks on Dara, the presumed successor to Jahangir. Whatever gratitude Aurangzeb felt toward Roshanara eventually dissipated, and over time, he grew closer to his sister Jahanara, who had sided with Dara in the ascension fight. But his relations with both women were complicated by one crucial edict, dating back to the time of Akbar: sisters of the Grand Mughal were not allowed to marry, for fear of producing offspring that might challenge the already fraught line of royal succession. According to multiple contemporary accounts, Aurangzeb expended a significant amount of effort keeping his two sisters from engaging in romantic and sexual relationships with men. In one account written by the French traveler François Bernier, Aurangzeb invited one of Jahanara's lovers to meet with him in his chambers, and offered the young man a betel nut as a gesture of hospitality. "Little did the unhappy lover imagine that he had received poison from the hand of the smiling monarch," Bernier wrote. "He died before he could reach home." In his *Memoirs of the Mughal Court*, the Italian Niccolao Manucci claims that Roshanara kept "nine youths in her quarters for her pleasure." When Roshanara's secret was uncovered, Aurangzeb had the nine men "destroyed in less than a month by various secret tortures." According to Manucci, the Grand Mughal went on to poison Roshanara herself for her offenses.

The rogue sex lives of the Mughal princesses made for salacious travelogues, but the daily reality of Mughal court society was one of patriarchal oppression. That reality was dominated by the institution of the harem, the gilded prison containing as many as five thousand women: the Mughal's wives and concubines, supported by their extended family of mothers and grandmothers, sisters and aunts, attended by ladies-in-waiting and female slaves, guarded by eunuchs. The elite women within the harem—the wives and concubines and their direct relatives—had an almost schizophrenic existence: they enjoyed standards of living unrivaled by almost any society on earth, while at the same time possessing almost no personal liberty. "These ladies lived in grand apartments luxuriously furnished, with lovely gardens, fountains, tanks and water channels attached to them," writes Soma Mukherjee, in her history of the Mughal princesses. "They wore beautiful and expensive clothes made from the finest material and adorned themselves with jewelry from head to toe." But their contact with the outside world was heavily regulated; on their rare excursions outside the harem, their faces were veiled in observance of purdah. Marriages were forced upon them, while their husbands were free to accumulate as many wives and concubines as they pleased. They were pampered aristocrats in the wealthiest society on the planet, and they were tightly controlled captives.

The fact that so many women in Aurangzeb's court were allowed to travel not just outside the harem's walls but all the way to modern-day Saudi Arabia tells us something about the importance of Islam inside the harem culture. Rigorous study of the Quran was an obligatory part of every Mughal princess's education. Aurangzeb's eldest daughter, Princess Zeb-un-Nissa, had memorized the entire Quran by the time she was seven years old. Their religious customs would occasionally warrant travel outside the harem to visit to a shrine or other holy places. But for a young woman in the court of Aurangzeb, the pilgrimage to Mecca would have been, by far, the longest journey of her life, and the only one outside the boundaries of the Universe Conqueror's dominion.

What would that experience have been like behind the veil of the Mughal princess, whoever she was? She and her sisters in the harem lived truly extraordinary lives, certainly compared to the vast bulk of society at the time: with so much wealth yet so little freedom, with the jewelry and the fountains and the sexual chains. But almost all our accounts of them come from the outside—most frequently through the observations of true outsiders like Manucci and Bernier. There were no Samuel Pepyses or Anne Franks among the Mughal princesses, or at least none whose words survived into the historical record: no diarist who could leave behind an honest record of what it felt like to be a woman in the Grand Mughal's court. It is obvious to our modern eyes that the harem was an institution of oppression, patriarchy rendered into stone. But would it have seemed like oppression to a woman who had been assiduously sheltered her entire life from any alternate reality? Or were there secret radicals in the harem, women who dreamed of a different way of organizing society? Had some of them wondered—rightly or wrongly—whether Europe itself offered a more appealing model, with its unveiled women, its monogamous marriage customs, its occasional female heads of state?

How you happen to answer those questions changes the way you think about the events that would transpire in September 1695, on the tropical waters of the Indian Ocean. It changes the way you think about Henry Every himself. Surely the Indian princess sailing back toward Surat must have understood that men had been granted far more liberty and power under the laws and conventions of the Mughal regime. Whether she had a language of oppression to describe that reality, she must have felt its sting, its reduced possibilities; she must have heard about—if not experienced directly—its ritualized rape. It seems plausible, at the very least, that those experiences would have left some kind of scar. The question is: Was that scar deep enough to make her want to escape?

Part Three

THE
HEIST

18

THE *FATH MAHMAMADI*

The Indian Ocean, west of Cape St. John
September 7, 1695

For more than a month, the newly formed pirate armada waited for monsoon season to arise, bringing the southwest winds that would carry the merchant ships back to Surat through the strait of Bab-el-Mandeb. With temperatures regularly climbing above a hundred degrees, there was no escaping the desert heat on Perim. (Even the water in the island's small natural harbor would have been in the low nineties at the height of summer.) The long wait made the men increasingly concerned that Every's plan was ill-conceived. "After they had lain there some time," John Dann would later recall, "they were apprehensive the Moors ships would not come down from Mocha, so they sent a pinnace thither, which took two Boats. They brought away two men, which told them the ships must come down." (Without a Suez Canal on the other end of the Red Sea, the merchant ships had nowhere else to go.) A few days later, they received word that the "Moors ships" were, at long last, headed toward the strait.

But when the first merchant vessels finally appeared on the last

Saturday of August, it initially seemed as though Every and his men had made a catastrophic error. Somehow, after a year's worth of preparation for this very moment, a convoy of roughly twenty-five merchant ships returning from Mocha managed to slip through the strait in the cover of night without the pirates detecting them. (No record exists of the punishment doled out to the multiple lookouts on watch that night, but presumably it was a severe one.) Only after the pirates captured a much smaller vessel the next morning did they realize their disastrous oversight.

According to Middleton, the pirates "consulted whether they should follow them or stay there." It was a difficult choice. The Indian ships had a near insurmountable head start, and one of the pirate ships, the *Dolphin*, was taking on water. But Every knew he had at his command the fastest vessel in the Indian Ocean. If any boat could overtake the Indian convoy, it was the *Fancy*.

After a quick consultation, the men decide to give chase, but with a streamlined fleet. The sixty men aboard the *Dolphin* were transferred to the *Fancy*, and the *Dolphin* was burned and sunk, on account of her being what John Dann later called an "ill sailor." It was clear that the *Pearl* had almost as meager a chance of keeping up with the *Fancy*, so the men lashed a rope between the two vessels and the fleet set sail toward the long-vanished Indian ships. Even lugging the *Pearl* behind her, the *Fancy* was the fastest ship in the convoy. Only the *Portsmouth Adventure* "kept them company," as Dann recalled. The *Susanna* would eventually catch up to the pirate fleet, but Thomas Tew's *Amity* drifted back beyond the horizon and lost contact with the other ships. Tew might have been the most celebrated Red Sea man at that moment in history, but the *Amity* was no match for the *Fancy* in the water.

For days, the pirate lookouts scanned the horizon for signs of their prey. Every steered the *Fancy* east-northeast, out of the Gulf of

Aden into the Arabian Sea. Without a ship to follow by sight, Every set a course for Surat and Bombay, assuming that one of the two ports was the treasure fleet's ultimate destination. Ten days passed with no sign of their target. The men were hungry and restless; their long wait on Perim had left them short on provisions. Their prospects grew even more depressing when, on the tenth day of pursuit, the lookout spotted land for the first time: the distant outline of Cape St. John, north of Bombay.

Those ten days and nights must have been agonizing ones for Every. He had traveled six thousand miles, assembled a battalion of four hundred men, and settled into the ideal spot to poach the most valuable treasure ships in the world—and then he had let them slip out of his sight. By the end of those ten days, with the early monsoon season winds rising and Cape St. John now visible, it was entirely possible that the ships had made it safely to harbor and were already unloading their goods beyond Every's reach.

But on the seventh of September, at long last, their luck improved. They caught sight of a handful of ships that had broken off from the main convoy. The largest of them was the heavily armed merchant ship *Fath Mahmamadi*, owned by one of the wealthiest traders in India, Abdul Ghaffar. A contemporary of Ghaffar's claimed that the merchant "drove a trade equal to the English East-India Company, for I have known him to fit out in a year, above twenty sail of ships, between 300 and 800 tons." For the first time since Every had concocted his scheme more than a year before, he had an actual Moor's treasure ship in his sights. Every quickly commanded his crew to sail ahead of the convoy, and then anchor overnight and wait for them— a risky strategy given the nighttime lapse that had destroyed their plans in the strait of Bab-el-Mandeb.

At dawn, a heavy mist hung over the water, restricting their visibility. The men stared out into the gray vapor, listening and looking for any signs of the Indian convoy. The wait did not last long. Within a matter of minutes the dark outline of the *Fath Mahmamadi* emerged from the fog, passing "within about a pistol shot of the *Fancy*." Dann later described her as being between "2 and 300 tons, with six guns." Every ordered the men to let loose with a broadside against the ship. The *Fath Mahmamadi* responded with three rounds that somehow did no damage to Every's ships, and then—amazingly—struck her colors in surrender. The ship was theirs.

On board the vessel, the pirates found significant reserves of silver and gold, worth upward of £60,000, the equivalent of around $5,000,000 today. It was a fraction of what Thomas Tew had stolen two years before, but it was certainly more wealth concentrated in one place than Every had ever seen in his lifetime. Divided among the three hundred or so men remaining, the treasure amounted to roughly three times the salaries that the Spanish Expedition had promised them for two years' worth of work. As captain, of course, Every had an additional share, meaning he probably walked away from the *Fath Mahmamadi* with enough money to live comfortably on land for half a decade, if not more. Abdul Ghaffar's treasure represented a life-altering influx of wealth to Every and his men. But Every would have been able to calculate at first sight of the treasure that it was not enough for him to retire from the game altogether.

Every had only forty-eight hours to savor his triumph. A detachment of men took command of the *Fath Mahmamadi*, and the fleet sailed farther east toward the coast. While anchored off the Cape of St. John on the tenth of September, the lookout sounded another alert. The sails of a far more formidable boat had appeared on the horizon. It was the *Gunsway*, making her final approach toward Surat. Within minutes, the *Fancy* was in pursuit again, under full sail.

The *Gunsway* must have made an intimidating first impression, once Every had sailed close enough to get an accurate assessment of her. A ship large enough to carry a thousand passengers—compared to the two hundred or so men crowded together on the *Fancy*—would have towered over the pirate vessels. Not only did the *Gunsway* have far more guns to defend herself—eighty cannons, hundreds of muskets—but its marksmen would be firing down onto the decks of the smaller ships, giving them a clear advantage in terms of their angle of attack.

Henry Every had made a thousand decisions over the fifteen months that had passed since the mutiny: when to careen his ship, how long to wait in Madagascar, whether to challenge Thomas Tew for the admiralty of the pirate fleet. But those first moments sizing up the full measure of the *Gunsway*—the threat and the opportunity—presented Every with a truly momentous choice. He had just pulled off a heist that gave him in a matter of hours five years of wages in one zero-fatality operation. He could walk away from the *Fath* raid one of the most successful pirates of his age, and if he was lucky, he might well encounter a few other strays from the Muslim convoy that were as easily overpowered. And clearly the *Gunsway* would not surrender as quickly. Every had lost half his armada, and one of his remaining ships was so sluggish that it had made more sense to tow her out of the Gulf of Aden than let her sail under her own powers. A betting man would have to assume that Every would lose in a direct challenge to the *Gunsway*.

Those calculations might well have played out differently if Every had known what had happened no more than few hundred miles away from him. After drifting behind the *Fancy*, Thomas Tew and the men of the *Amity* had their own separate run-in with the *Fath Mahmamadi*, engaging the Indian ship in battle. The attack ended in catastrophe for the *Amity*. During the firefight, a cannonball sliced across Tew's belly, effectively disemboweling him. According to one report, he died holding his lower intestines in his hands: "When he dropp'd it struck such a Terror in his men, that they suffered themselves to be taken, without making Resistance." A number of the pirates were captured and taken away as prisoners.

If the violence and gore of Tew's death was shocking, so was the narrative of his final years. Here was a man who, thanks to his 1693 heist, could have chosen to retire to a life as a landed gentleman in Rhode Island, an American Francis Drake—and never lift a finger performing hard labor again. And yet despite that available future, he had ended his life on the other side of the world, staring at the sky above the Gulf of Aden, holding his bowels in his hands as he bled out on the *Amity*'s deck.

There is a tantalizing possibility that Every *did* know about Tew's horrific death. Some accounts of the battles in the Indian Ocean that late summer claim the *Amity* had had its fatal encounter with the *Fath* just a few days before the *Fancy* had overtaken her. (This chronology offers one potential explanation for why the *Fath* was so quick to surrender: she had been damaged in the exchange with the *Amity*, and lacked the firepower for another fight.) In this scenario, when Every boarded the *Fath Mahmamadi* to inspect its reserves of silver and gold, he would have discovered prisoners captured from the *Amity*, who would no doubt have informed him of his fellow captain's shocking demise.

But the more likely scenario is that Every weighed the odds of an attack on the *Gunsway* oblivious to Thomas Tew's death. Why would a man who had played his cards so carefully over the past months— artfully trying to carve out a place for himself on the right side of British law, waiting patiently for the monsoon season to bring the Moor ships into his trap—make such a high-risk assault on a ship that seemed destined to blow him out of the water? Perhaps Every had been deceptive in the enigmatic closing lines of his letter;

perhaps his "hunger" for treasure was every bit as excessive as his men's. Perhaps he feared that he himself would be the victim of a mutiny if he didn't make a run at such a tremendous prize. Or perhaps he simply trusted that the raw speed of the *Fancy* would enable him to mount an attack on the *Gunsway* and retreat quickly if things turned ugly. *She sails so hard now*, the East India Company correspondent had warned a few months before, *that she fears not who follows her.*

The same letter had predicted "infinite clamors" at Surat if Every were left unchecked. The decision to make a direct assault on the *Gunsway* would turn that seemingly hyperbolic warning into prophesy, only with "clamors" that extended far beyond Surat itself. Every and his men were hungry; they were fearless. And they had exceeding treasure in their crosshairs at last.

19

EXCEEDING TREASURE

The Indian Ocean, west of Surat
September 11, 1695

Imagine taking a bird's-eye view over the coastal waters of the Indian Ocean on that late summer day. To the east you can make out the shipyards and factories lining the Tapti River at Surat. Somewhere in those buildings, Samuel Annesley is reviewing inventories or writing reports to be sent back to London, with no knowledge of the catastrophe that is about to befall him and his colleagues. In the waters outside Surat, the advance guard of the Mecca convoy is making its way into the harbor, their passengers weary from the three-month journey, but relieved to have made it back home without any confrontations with the Red Sea Men. And then farther out at sea, maybe a hundred miles from the coastline, two ships stand out: one twice the size of anything else in the water, proceeding slowly toward the harbor at Surat; the other surging across the waves, sails unfurled, its crew scrambling on the decks to prepare the guns for the coming battle.

Ancient history is always colliding with the present in the most literal sense: our genes, our language, our culture all stamp the present moment with the imprint of the distant past. But this scene in the Indian Ocean in 1695 is a different kind of nexus, one of those rare moments where multiple long arcs collide in spectacular fashion: the Indian wealth that had first started to accumulate with those cotton fabrics invented in the centuries before the birth of Christ; the itinerary of the pilgrims defined by Muhammad's own trek to Mecca a thousand years before (and perhaps by Abraham's before that); the vast power of Aurangzeb, handed down through the generations of Muslims ruling the subcontinent; the faltering fortunes of the East India Company, struggling to maintain its footholds in Surat and Bombay; the long history of piracy and its radical egalitarianism. Tell any of those different stories on its own, and the events of September 1695 would still register prominently on each timeline.

What unites all those different threads? Two hundred men, six thousand miles from home, clustered together on a ship, low on supplies, intent on making their fortune.

There is general agreement about three things that transpire once the *Fancy* pulls close enough to the *Gunsway* to mount an attack. First, a cannon explodes on the decks of the Mughal ship, killing a half dozen men, gravely wounding others, and creating a general scene of chaos and destruction just as the *Gunsway* is preparing to fire back on Every's ship. Second, one of Every's first volleys turns out to be a spectacularly lucky one, striking the mainmast of the *Gunsway* and causing the mainsail and all of its rigging to collapse, crippling the ship and adding to the existing chaos triggered by the malfunctioning cannon.

The third undisputed fact is this: by the end of the encounter, Every and his men capture a fortune worthy of the *Gunsway*'s name—"exceeding treasure." In the immense hull of the ship, they find astounding quantities of gold and silver, along with jewels, ivory, myrrh, frankincense, saffron, and other delights. From the time news broke of Every's plunder, a debate has raged over exactly how much the pirate managed to steal from the Indian treasure ship. Captain Johnson, in his *General History of the Pyrates*, expressed the difficulty of calculating the amount: "It is known that the Eastern People travel with the utmost Magnificence, so that they had with them all their Slaves and Attendants, their rich Habits and Jewels, with Vessels of Gold and Silver, and great Sums of Money to defray the Charges of their Journey by Land; wherefore the Plunder got by this Prize, is not easily computed." Some estimates suggest that the treasure was worth around £200,000, roughly $20,000,000 in today's currency. The East India Company later priced the stolen goods at three times that number. However you calculate it, Every's plundering of the *Gunsway* ranks as one of the most lucrative heists in the history of crime.

But once you look behind those three undisputed facts—the exploding cannon, the collapsed mainmast, and a plunder of historic proportions—the story of Every and the *Gunsway* bifurcates into two competing narratives. In one account—the narrative that would be sung by the balladmongers and sensationalized by the pamphleteers back in London for decades to come—the pirates board the ship and engage the Indian warriors in hand-to-hand combat for two hours before overcoming them. With the ship now under their control, the men discover, to their astonishment, dozens of Muslim women in hijabs cowering belowdecks. The emeralds and diamonds that adorn their garments signal that they are members of Aurangzeb's court.

One of those veiled faces, according to this account, belongs to Aurangzeb's granddaughter. Somehow the pirates determine her true identity, and bring her, sobbing in terror, before Every himself. According to Van Broeck's 1709 narrative, "the Captain no sooner beheld the Lady in Tears, but melted into Compassion." What follows from that moment is an Indian Ocean rendition of the

Pocahontas story: the Westerner entranced by the "native's" exotic beauty, finding love at the end of a dangerous voyage. (The twist in this story, of course, is that the "native" is far more affluent than the Westerner.) The story goes that Every proposes to the Mughal princess on the spot, having at long last found "something more pleasing than jewels." Married in front of a Muslim cleric, "the happy newly-weds were said to have spent the whole trip back to Madagascar engaged in conjugal bliss."

Daniel Defoe's rendition of the story, recounted in his 1720 book *The King of Pirates* as a first-person narrative in Every's voice, gives a similar account of the Mughal princess's appearance in Every's quarters (though Defoe upgrades her to a queen): "Such a Sight of Glory and Misery was never seen by Buccaneer before; the Queen (for such she was to have been) was all in Gold and Silver, but frighted; and crying, and at the Sight of me she appear'd trembling, and just as if she was going to die. She sate on the Side of a kind of a Bed like a Couch with no Canopy over it, or any Covering, only made to lie down upon; she was, in a Manner, cover'd with Diamonds, and I, like a true Pirate, soon let her see that I had more Mind to the Jewels than to the Lady." As in Van Broeck's account, Every treats the Mughal queen with respect, but in the Defoe version, a romance between the two fails to blossom. Apparently the pirate captain has designs—honorable, amorous ones—on one of her ladies in waiting: "There was one of her Ladies who I found much more agreeable to me, and who I was afterwards something free with, but not even with her either by Force, or by Way of Ravishing."

The love story of the Devonshire pirate and his Mughal bride seems, to the modern eye, implausible at best. Annesley's biographer records that Every "carried off as captive a young Mohammedan lady of good family who was proceeding to her home after performing the pilgrimage to Mecca." But the key word in that sentence is surely "captive," a word not conventionally associated with conjugal bliss. Even in some of the more sentimental accounts of Every's courtship, you can hear the doubts creeping into the narrator's voice. "I have heard that it has been reported in England that I ravish'd this Lady," Defoe's narrator writes, "and then used her most barbarously; but they wrong me, for I never offer'd any Thing of that Kind to her, I assure you." Defoe has Every defending the actions of his crew as well: "If any of the Princess's Women were lain with, said I, on Board the other Ship, as I believe most of them were, yet it was done with their own Consent and good Will, and no otherwise; and they were all dismiss'd afterwards, without so much as being put in Fear or Apprehensions of Life or Honour."

Van Broeck offers a comparable defense of Every's honor: "Instead of ravishing the Princess, which some Accounts have made mention of, [he] pay'd the Respect that was due to her high Birth, took her and her Attendance into his own Ship, and after despoiling the Vessel of all its Wealth suffer'd it and its Crew to steer on to their intended Port." Yes, he and his crew might have "despoiled the Vessel" of all its wealth, but at least Every had given the princess the respect she deserved. No matter what those other "accounts" might have you think.

What were those mysterious accounts? The fact that there was a challenge to the popular narrative of Every as a lovestruck pirate—not only respecting the princess by asking for her hand in marriage, but also agreeing to a service officiated by a Muslim cleric—was itself a meaningful development in the relationship between England and India. European sailors—whether pirates, merchants, or naval officers—had been committing barbaric crimes in remote parts of the world for at least two centuries, from Drake's bloody march across the harbor towns of Central America to the genocide perpetuated by the Dutch in the Spice Islands of Indonesia. But stories of those atrocities rarely made it back to the European capitals, to call into question the moral righteousness of those bold explorers. Back home, they were heroes, not mass killers. But Henry Every and his men would not be able to outrun the infamy of their actions. In the attack on the *Gunsway*, the victims produced their own rival version of the events, a counternarrative that was far less forgiving.

The fact that this narrative emerged to challenge the soft-focus, idealized account of the *Gunsway* affair was partly a reflection of the power dynamics that existed between India and England at the time. The thirteen thousand Bandanese murdered by the Dutch off the Indonesian mainland in the early 1600s had no warships or ambassadors or clerks to record and protest the atrocities committed against them. But Henry Every was taking on a ship that belonged to the wealthiest man in the world, a man who commanded an immense state apparatus that rivaled any of the "civilized" governments of Europe.

The second narrative emerged for another reason, too—one of those fortuitous moments where the perfect witness just happens to be at the right place at the right time. While Every was mounting his attack on the *Gunsway*, an emissary for the commander of the inland town of Rahiri had arrived in Surat to conduct some transactions that would later take him to Bombay. The emissary's name was Khafi Khan. Whether Khan successfully delivered the goods back to his employer is not known, but in Surat, he stumbled across something that would be far more meaningful: a story of abuse and murder so shocking that he immediately recognized its political significance. Khan was uniquely suited to make such a recognition because he was not just a courier; he also happened to be an aspiring historian, and would go on to great acclaim as the authoritative chronicler of Aurangzeb's regime. The Indian fleet had suffered several outrageously unlucky breaks in their confrontation with Every. But in Khafi Khan, fortune turned in their favor for once: a master storyteller landing in Surat just in time to intercept a ship bearing news of the crime of the century.

20

THE COUNTERNARRATIVE

The Indian Ocean, 50 miles west of Surat
September 11, 1695

Khafi Khan based his account of the *Gunsway* raid on direct testimony from its survivors, delivered after the Mughal ship finally made it to Surat, a week after Every had first spotted her. Early versions of Khan's narrative circulated through Aurangzeb's extended court network, eventually reaching the Universe Conqueror himself. They were ultimately published as part of Khan's broader history of the Mughal dynasty, a sweeping chronicle that extended from Akbar's reign to the then present-day rule of Aurangzeb himself. The son of a historian, Khan spent much of his life in the employ of Aurangzeb in various positions, serving as his eyes and ears as a kind of traveling court reporter. It was this occupation that gave his accounts of the *Gunsway* affair such immediacy. It also made his account more true to life than the thirdhand fantasies spun by the balladmongers. The grand romance of Every and the Mughal princess belonged to the tabloid hacks, telling stories based on indirect accounts. Khan was something else: a historian effectively functioning as journalist, interviewing witnesses who had been passengers on the *Gunsway* when Every's men boarded her.

Khan dives almost immediately to the question that Aurangzeb would have first asked: How did a ragged band of pirates conquer a ship at least three times their size and force? As an answer, he dutifully reports the improbable opening volley and the exploding cannon, but he offers up a third explanation, one that most European accounts would subsequently ignore: that the captain, a debauched aristocrat named Ibrahim Khan, had a crisis of nerve. After the collapse of the mainmast, Every's men clambered aboard the ship from both the port and starboard sides and engaged the Muslim warriors in a frenetic swordfight straight out of an Errol Flynn movie. Some of Aurangzeb's men fought off the invaders valiantly, but Ibrahim Khan seems to have lost his wits in the chaos of Every's attack:

> The Christians are not so bold in their use of the sword, and there were so many weapons on board the royal vessel that if the captain had made any resistance, they must have been defeated. But as soon as the English began to board Ibrahim Khan ran down into the hold. There were some Turki girls whom he had bought in Mocha as concubines for himself. He put turbans on their heads and swords into their hands, and incited them to fight. These fell into the hands of the enemy, who soon became perfect masters of the ship.

The story of Captain Khan's cowardice is so bizarre that it likely had some kernel of truth to it. It is sometimes retold with comic undertones: a hapless captain sending a posse of women in drag to defend his ship. But the reality must have been terrifying for the women living through it. Imagine the experience from the vantage point of those Turkish "concubines"—sex slaves bought and paid for like so many barrels of coffee beans, imprisoned in the hold of a royal ship. Imagine lying there in the dark, hearing the deafening sound of the cannon exploding above you, smelling the burning planks and the gunpowder wafting belowdecks. And then, out of nowhere, your captor—the man who has almost certainly been raping you repeatedly for the past month at sea—dashes into your cramped quarters, drapes you in a turban, and presses a cutlass into your hands. Imagine staggering out into the melee on the main deck, trying to make sense of the anarchy that surrounds you.

As horrific as it must have been, the Muslim captain's actions would pale beside the offenses the English were about to commit. At the end of combat, twenty-five men (and, presumably, women) were dead on the *Gunsway* side, with almost as many gravely wounded. In all the chaos, Every had not lost a single man. (Justifying this devastating defeat, Captain Khan later spun his own account, claiming that Every had attacked his ship with an army of 1,200 pirates.) The pirates immediately began searching the ship for treasure. Some of the immense wealth aboard the *Gunsway* was easy to find: the piles of gold and silver bullion, the barrels of valuable spices. Dozens of Every's men began lugging the booty back to the *Fancy* under the watchful eye of the quartermaster, Joseph Dawson, whose job it would be to divide up the spoils once they had been fully extracted. But the pirates knew an imperial ship like the *Gunsway* would almost certainly have additional valuables hidden somewhere on board. To determine the location of that concealed booty, the men reverted to the standard practice of pirate inquisition: torture.

The exact techniques used by the pirates to compel the *Gunsway*'s officers to reveal the ship's hidden treasure were not recorded in any account of the 1695 attack. But the subsequent outrage suggests they were severe ones. As a guide, we can only use firsthand and reported

accounts of other pirates attempting to extract the same information
from other crews. According to a dispatch in the *American Weekly
Mercury*, the eighteenth-century pirate Edward Low—whose articles
of agreement are one of the four main pirate constitutions to have
survived to the modern age—"cut and whipped some and others they
burnt with Matches between their Fingers to the bone to make them
confess where their Money was." The Dutch journalist Alexandre
Exquemelin described the technique of "woolding" used to extract
the location of treasure from uncooperative sources: "They strappa-
do'd him until both his arms were entirely dislocated, then knotted
the cord so tight round the forehead that his eyes bulged out, big as
eggs. Since he still would not admit where the coffer was, they hung
him up by his male parts, while one struck him, another sliced off his
nose, yet another an ear, and another scorched him with fire."

But even these acts are secondary charges, in Khan's account.
The real offense centers on the female passengers of the *Gunsway*.
The fury of battle, the fifteen months at sea in a bubble of unchecked
masculinity, the anti-Muslim bigotry that Every's men had exhibited
destroying the Mosque in Maydh, and the sudden discovery that the
ship they have just boarded contained dozens of women, some of
them adorned with jewels more valuable than the pirates' collective
net worth—all these forces coalesced to trigger an eruption of sexual
violence that lasted for days. The key passage from Khan—echoed in
a long chain of court reports, letters, and word-of-mouth gossip—
tells the story in two sentences:

> When [the pirates] had laden their ship, they brought the
> royal ship to shore near one of their settlements, and busied
> themselves for a week searching for plunder, stripping the
> men, and dishonoring the women, both old and young. . . .
> Several honorable women, when they found an opportunity,

threw themselves into the sea, to preserve their chastity, and some others killed themselves with knives and daggers.

There is a strange reluctance in the literature of piracy to center the camera on these kinds of offenses—strange because that literature is otherwise happy to feed you the gore and terror of the pirate's life in such intimate detail. If you want to read about Thomas Tew dying on board the *Amity*, holding his small intestines in his hand, or Edward Low ripping the beating heart out of a prisoner, there are a thousand pages in the archives that will give you that experience, uncensored. Gang rape, on the other hand, gets condensed down to a euphemism: *Then the men dishonored the women.*

A similar description of the events appears in a private correspondence filed weeks later by John Gayer of the East India Company, after he met with Khafi Khan in Bombay. Echoing Khan's narrative, Gayer reported:

> It is certain the pirates, which these people affirm were all English, did do very barbarously by the people of the Gunsway and Abdul Gofor's Ship, to make them confess where their money was, and there happened to be a great Umbraw's wife (as we hear) related to the king, returning from her pilgrimage to Mecca, in her old age. She they abused very much, and forced several other women, which caused one person of quality, his wife and nurse, to kill themselves to prevent the husbands seeing them (and their being) ravished.

Ravish, dishonor: To the modern ear, the words sound too mannered, too circumspect for the crime they describe. We should not mince words: Every's men were rapists of the worst order.

Khan's language displays the same discretion as Gayer's, no doubt to avoid offending Aurangzeb's fundamentalist beliefs. A religious culture that insists on veils for its women will not take kindly to vivid descriptions of the sexual crimes committed against them. But Khan stressed in both his own writing and his testimony to Gayer one shocking fact that conveys, by transitivity, just how offensive the English assault on the *Gunsway* was: members of Aurangzeb's court had deliberately chosen to stab themselves in the heart or throw themselves overboard to avoid the "dishonor" that Every's men would otherwise subject them to. Whatever happened during those long days and nights while Henry Every's men terrorized the women of the *Gunsway*, it was so horrendous that suicide seemed the better option. Years later, in his dying words, John Sparkes would claim that "the inhuman treatment and merciless tortures inflicted on the poor Indians and their women still affects my soul."

However oblique, Khan's description of the "dishonor" those women were subjected to made an indelible mark on the story. The crimes of Henry Every and his men would be numerous: mutiny, murder, torture, theft. But after Khan, the rape of the *Gunsway* women would always loom large in the charges leveled against them. The pamphleteers and balladmongers would continue to sing the praises of Every's chivalrous behavior, but among the authorities that eventually became bound up in the case—Aurangzeb's court, the East India Company, and the English government—the gang rape aboard the *Gunsway* went unchallenged as a central fact in the case.

Where in all of this depravity can we find Aurangzeb's granddaughter? Gayer had made the vague allusion to a "great Umbraw's wife related to the king" as well as "one person of quality" who committed suicide. ("Umbraw" is an anglicized version of the Urdu word *umara*, meaning a grandee of the royal court.) Certainly there were relatives of the Grand Mughal on board the vessel when Every attacked her; it seems likely from Gayer's account that at least some of them were "abused" by the pirates, or driven to suicide by the threat of that abuse. Was one of them a young princess, traveling to Mecca as a religious pilgrim, who was captured and brought aboard the *Fancy* and presented to Every himself? And what transpired in that encounter, if it did happen? Subsequent events and testimony make it seem likely that some kind of interaction did take place between Henry Every and one of Aurangzeb's relatives—whether a granddaughter or some more distant connection. But was it merely a continuation of the sexual violence erupting on the *Gunsway*, or could it have somehow been closer to the legend that evolved over the subsequent decades, the unlikely multicultural romance of the English pirate and his Muslim bride?

According to Philip Middleton's testimony, Henry Every never boarded the *Gunsway* during those violent days in September. But he must have sensed that his men had crossed a line from thieves into something far more repugnant: torturers, rapists, enemies of all mankind. A sailor with Every's experience and cunning would have recognized immediately the consequences that those actions would trigger, once word of them finally reached the mainland. The crew of the *Fancy* were in new waters now. Every's prediction that unimaginable wealth was to be found in the treasure fleets of the Red Sea had been validated, as had his dark prophesy that his men would "exceed his desire" given the chance. The question was whether the second prophesy coming true would undo the triumph of the first.

Against extraordinary odds, Henry Every had made his fortune. But he must have realized, listening to the screams echoing across the water from the *Gunsway*, that his men's actions had now made him something else: the world's most wanted man.

21

VENGEANCE

Surat, India
Mid-September 1695

I t only took a matter of hours after the battered crew of the *Fath Mahmamadi* pulled into the Surat harbor for word to spread through town that Abdul Ghaffar's ship had been attacked by English pirates, with "severall of their Men killed in fight, and others barbarously used." In the chief factor's quarters overlooking the dockyards, Samuel Annesley would have immediately grasped that the news of more British piracy did not bode well for the East India Company. Many Surat residents already suspected that the company was supplementing its trade revenue by stealing directly from Indian merchant ships, through some kind of quiet partnership with the pirates. That hearsay turned into a direct accusation when Abdul Ghaffar, Surat's wealthiest merchant and the owner of the *Fath Mahmamadi*, learned that English pirates had plundered his ship. As Arnold Wright, Annesley's biographer, puts it: "The avenging finger of Abdul Guffor was pointed toward Annesley and his colleagues as the real authors of the crime."

By September 12, a mob of enraged locals had gathered at the gates of the East India Company factory, demanding vengeance for the company's abuses. Circulating among them was Khafi Khan, taking notes and interviewing the crew of the *Fath*, gathering evidence for the report he would eventually send back to Aurangzeb. At first, Annesley took the gathering protest in stride. He ordered the gates of the factory closed, assuming he could wait out the storm. "He knew the capabilities of the place for defence," Wright explained, "and had no misgivings as to the outcome of a fight between the well-armed inmates and the miscellaneous crowd of ruffians which the bazaars of Surat were able to furnish in times of disorder."

A few hours later, the Mughal military commander at Surat, Usher Beg, arrived at the gates "with a troop of horses clattering at his heels." He secured admission to the factory by claiming that he had a message from Surat's governor, but the message turned out to be a ruse. Instead, Beg was there to put Annesley and his men under house arrest while the authorities investigated the plundering of the *Fath*. Commander Beg claimed he and his troop had been dispatched to the factory to protect the English from the mob outside the gates, but Annesley suspected that something more sinister was afoot. Yet Annesley had some reason to take Beg at his word; the Englishman had enjoyed friendly relations with the *mutassaddi* (governor) of Surat, I'timad Khan, and had kept the company on his good side with a reliable stream of bribes over the years. It was better, Annesley figured, to accept the house arrest and the protection of the Mughal guard and allow the fury in the streets to subside.

With the company men secured behind the factory walls, the agitators on the street—led by the town's elder clerics—made a direct petition in front of the governor, demanding that Annesley and other key agents of the company be executed for their alleged involvement in the crime. The governor listened patiently to the long list of grievances, but declined to pass judgment. He did, however, promise to convey the facts of the case to Aurangzeb, and deliver whatever punishment the Grand Mughal considered appropriate. Just as Annesley had resigned himself to a few days of house arrest, the governor likely assumed he could buy himself—and the company—some time by deferring to Aurangzeb's wishes. It would take weeks for an account of the *Fath Mahmamadi* piracy to reach the court at Delhi; by that time, the whole affair would hopefully have blown over, and his lucrative partnership with the East India Company would be back in business. If Aurangzeb sided with the protestors, I'timad Khan could always point to the potential lost revenue that would result from evicting the company altogether. Even if you accepted the improbable premise that the pirates were hired thugs working for the East India Company, Aurangzeb was gaining more from the tariffs and bribes that the company paid to the Mughal authorities as a cost of doing business in Surat than whatever he was losing to piracy.

That financial calculation collapsed just two days later, when the *Gunsway* and its traumatized survivors anchored in Surat. "The capture of the Imperial pilgrim ship in Mohammedan eyes was more than a crime," Wright observed. "It was sacrilege." The British were not just guilty of stealing from wealthy merchants; they had committed appalling acts of sexual violence against the women of Aurangzeb's extended court, women who were taking part in the most sacred journey in the Muslim faith. It was hard to imagine a crime better engineered to infuriate Aurangzeb. Henry Every—wittingly or not—had transgressed the most cherished of the Universe Conqueror's possessions: his fortune, his faith, and his women.

During his conversation with the victims and survivors, some of whom were personal acquaintances, Khafi Khan heard a troubling refrain. In the frenzy of their attacks, some of the British were heard to say that they were taking revenge for the siege of Bombay, suggesting that they had themselves been imprisoned during that long standoff five years before. By definition that would have made them members of the extended family of the East India Company, if not direct employees. Those accounts would prove to be a key piece of evidence in the case. Nothing we know about the members of Every's crew suggests that any of them were participants in the siege of Bombay. But the truth is we know very little about Every's crew, almost as little as we know about Every himself. It is entirely possible that some of them men who signed up for the Spanish Expedition had in fact worked for the company and had suffered through the siege. Or perhaps invoking the siege was the sort of anti-Muslim slur an Englishman at that time might have uttered, a seventeenth-century version of "Remember the Alamo!" Whatever the reality, the story did not paint the company in a sympathetic light. If Gayer and Annesley had helped plot this attack as retaliation for the siege of Bombay, it was not just sacrilege. It was an act of war.

The protests at the governor's mansion took on a fever pitch. "The town is so defiled that no prayer can be offered up acceptable to God til Justice is done," Abdul Ghaffar thundered. As Ghaffar and the clerics relayed the astonishing facts of the case to I'timad Khan, the governor recognized that the stakes had changed, for good. This was a storm that would not quickly blow over. The mob was gathering outside *his* gates now, not just the gates of the factory. If he failed to punish the English sufficiently, his own life could be in danger. Without waiting for guidance from Delhi, Governor Khan ordered that every Englishman in Surat be rounded up and imprisoned in the East India Company factory. Annesley and his colleagues were chained in heavy iron, "like a company of Doggs." For a stretch of time, the English were deprived "the libertye of a Penn and Ink," their communication with the outside world cut off entirely.

After an initial blackout, the correspondence with Bombay Castle was restored. (Annesley developed a secret code in his messages, convinced that his captors were reading his exchanges with Gayer.) By Annesley's own account, it was a miserable existence, waiting in chains for the wrath or mercy of Aurangzeb's judgment, knowing that at any minute the mob could storm the gates and exact their revenge directly. "It is needless to write of the indignities, slavish usages and tyrannical insultings wee hourly bear day and night," he wrote to Gayer, once his "Penn" had been restored, "and to expatiate on so hateful a subject woud no wayes redress or alleviate our sufferings."

To his captors, Annesley continually made the case that it ran against the company's interests to sponsor piracy when so much of their revenue depended on the good grace of their trading partners in Surat and Bombay. "For nine years past," he wrote to Governor Khan, "[there] have been the same false aspersions on us and all along wee have at last merchants and not pyrates. If wee were the latter, wouldst wee live amongst them and so many 100,000 rupees' worth of goods to the City?" Privately, I'timad Khan was sympathetic. Publicly, his hands were tied. He didn't dare liberate prisoners that had at least a circumstantial connection to the crew of the *Fancy* before Aurangzeb weighed in on the case.

News of the *Gunsway* affair reached Delhi sometime in the early fall of 1695. It may have been delivered directly to Aurangzeb by Khafi Khan himself. Accompanying the narrative of British atrocities were two key pieces of evidence, designed to enrage the Grand Mughal: survivors of the *Gunsway* attack, some of them potentially relatives of Aurangzeb, testifying to the moral depravity of the British pirates; and coins that had been minted in Bombay, bearing the image of King William, evidence of the British thumbing their noses at the Universe Conqueror's sovereign power. The emissaries from Surat made a stern argument about the culpability of the company in the attacks on the *Gunsway* and the *Fath*. Piracy was not something the East India Company merely turned a blind eye to, the Surat contingent argued. Piracy was a key part of their business model. Khafi Khan had done the math: "The total revenue of Bombay, which is chiefly derived from betel-notes and coco-nuts, does not reach to two or three lacs of rupees. The profits of the commerce of these misbelievers . . . does not exceed twenty lacs of rupees. The balance of the money required for the maintenance of the English settlement is obtained by plundering the ships voyaging to the House of God, of which they take one or two every years."

Not surprisingly, the case against the British fell on receptive ears. (As Wright put it, "To such a fanatical and arrogant [ruler], the audacious crimes of Every were calculated to be as a spark introduced into a barrel of gunpowder.") Appalled by the sacrilegious acts of the English "infidels," Aurangzeb ordered his men to seize the assets of the Surat factory, and to prepare for an assault on Bombay Castle. The East India Company had tested the Universe Conqueror's patience one too many times. The attacks on the *Fath* and the *Gunsway* had given the lie to the long charade of the English as business partners with the Mughal empire. Their true colors had been revealed in Every's lawless deeds: the company was an invading force, threatening Aurangzeb's sovereign rule and desecrating his religious beliefs. It was time to expel them.

22

A COMPANY AT WAR

Bombay Castle, Bombay
Fall 1695

The *Gunsway* attack might reasonably have seemed an act of war to Aurangzeb, given the evidence presented to him by Khafi Khan and the other Surat emissaries. But if it was an act of war, it was a strange one, at least measured by contemporary definitions. Strictly speaking, the military conflict that threatened to erupt in the fall of 1695 was one between an empire and a corporation much more than it was a conflict between two sovereign nations. William III had not declared war against India in any formal sense. (He was too busy fighting the French in the closing acts of the Nine Years' War, and grieving the death of Queen Mary, who had succumbed to smallpox at the end of 1694.) Even if William III had intended to engage with India militarily, the simple fact of the matter was that the East India Company was far better situated to conduct such an operation. The company possessed richer and more reliable information networks connecting London to the subcontinent; they had a fleet already sailing in the region; and their headquarters in Bombay was, literally, a fort. The Royal Navy had its own fleet, of course, but in every other respect they were far less suited to wage war in Southeast Asia than the company was.

The modern mind—accustomed to the geopolitical structures that cohered in the centuries after Every—struggles to find an equivalent for this strange relationship between crown and corporation. Perhaps the best way to think about it is that England outsourced the problem—and the opportunity—of India to a private subcontractor. That private company was effectively given carte blanche to negotiate trade deals, engage in naval battles, acquire territory—all powers that are now reserved for nation-states, not private companies. If that distribution of authority seems strange to us now, it was not necessarily any more intelligible to the participants at the time, in part because the categories were so new. What was the proper role and responsibility of a multinational corporation in its dealings with a foreign power? No one really knew. This was a time, as the historian Philip Stern observes, when "national territorial states did not have a monopoly on political power and in which sovereignty was composite, incomplete, hybrid, layered, and overlapping."

A comparable ambiguity applied to Every and his men: Could you make a living as a pirate plundering Muslim treasure ships and still consider yourself within the sphere of legitimacy as an English citizen, the way Francis Drake had managed to do? This, too, was an open question in the fall of 1695. Yes, Henry Every had clearly run afoul of British law by stealing a ship that belonged to the Spanish Expedition investors, but they had arguably breached their contract with the crew by failing to compensate them for their labors. Every lacked a letter of marque, but his actions at sea, and his public letter to the British authorities promising not to attack British ships, suggested a man trying to define a space for himself and his crew at the very edges of legitimacy.

All of which meant that the key English participants in the

crisis—Every, Annesley, Gayer, the directors of the company back in London, even King William himself—were all probing the limits of their respective roles, because those roles had not been entirely defined yet. They were, effectively, helping to define these new institutions by exploring their boundary conditions. There were three distinct categories: pirates, corporations, nations. No one was quite sure where one began and the other ended. Much of the global crisis that Henry Every's actions provoked had roots in that underlying confusion.

From Aurangzeb's perspective, of course, those blurred boundaries presented fewer taxonomic challenges: pirates, company factors, kings—they were all Englishmen. But for the Englishmen living under Aurangzeb's reign—for Samuel Annesley and John Gayer most of all—the need to create a conceptual division between the pirates and the corporation, in Aurangzeb's mind at least, had become an existential one. There would be no more East India Company in India if the Grand Mughal could not be convinced that the distinction was meaningful. On October 12, Gayer composed a letter reporting back to London on the volatile events of the preceding month. He ended with this ominous line: "The Pyrates, being neglected of all hands, begin to grow formidable, and if some Course be nott taken to destroy them, they will yearly increase, having found their trade so beneficiall, and how soon the Companys servants, as well as their Trade, may be sacrificed to revenge the Quarrell of the Sufferers, they know not."

Back in London, the company was already facing a different kind of existential threat. Those intoxicating calico fabrics that India had introduced to the world so many centuries before were not just draping the bodies and sitting rooms of well-to-do urbanites. They were also undermining England's domestic wool trade. All across northern England, a de facto "Make England's Wool Business Great Again" movement arose, claiming that the hardworking native laborers were being undercut by the artisans of the subcontinent (and the middlemen who brought their foreign wares to British soil). "When the East India ships come in," they argued, "half our weavers pay." The charges also invariably featured a pronounced tone of sexual shaming. "Calico madams" were destroying one of Britain's most enduring industries through their suspiciously sensual embrace of cotton. Real Englishwomen, apparently, wore wool. It was a message that resonated even among the chattering classes back in London. The balladmongers spun songs about the calico madams; poems and pamphlets renounced them. ("None shall be thought / a more scandalous Slut / Than a tawdry Calico Madam," one of them declared.) Daniel Defoe described the craze for cotton as a "Disease in Trade . . . a Contagion, that if not stopped in the Beginning, will, like the Plague in Capital City, spread itself o'er the whole Nation."

By March 1696, the protectionist defenders of the wool industry—one of the first genuine labor movements in history—had convinced at least some members of Parliament that drastic measures were required. Already the more radical body, the House of Commons passed a bill banning the importation of "all wrought silks, Bengalls, dyed printed or stained calicoes of India." By that point, the share price of East India Company stock had lost half its value in just fifteen months. If the House of Lords ratified its own version of the bill, the weavers of the North Country would level almost as devastating a blow to the company as the one Aurangzeb was threatening to deliver. Take silks and dyed cottons off the balance sheet, and the East India Company would be out of business.

That was where the world's first joint-stock multinational corporation—now arguably the most powerful economic force on the planet, rivaled only by the economic power of national governments—found itself a few years shy of its hundredth birthday, facing existential threats at home *and* abroad. If England had out-sourced its relations with India to the East India Company, it was not clear, in those first months of 1696, how that strategy was going to play out. Perhaps powerful imperial states would come to realize—as Aurangzeb seemed on the verge of doing—that the capitalist trad-ers were undermining their authority and exploiting their economies both through parasitic trading practices and outright piracy. Perhaps the weavers would figure out a way to get the Tory House of Lords on their side, and the East India Company would collapse via an act of Parliament. Both were viable scenarios.

We should not push the alternate histories too far. The joint-stock multinational corporation was probably going to become a dominant organizational form, one way or another, however trou-blesome the anti-calico crowd or the pirates happened to be. But if there were a moment when its long-term survival was most at peril, it was probably during that last decade of the seventeenth century, thanks to the bribery scandal, the siege of Bombay and the Every affair in India, and the "calico madam" backlash at home. It was an inflection point, a stretch of history's river where small perturba-tions can determine which way it ultimately runs.

Confronting this crisis from within the fortifications at Bombay Castle, John Gayer wrote increasingly desperate reports back to Lon-don. The first made it to the company's headquarters on Leadenhall Street in East London in December 1695; three others followed in the coming months. If Every and his men were not apprehended and brought to justice, Gayer warned, the wrath of Aurangzeb would result in complete expulsion from the subcontinent, if not mass slaughter of the company's employees. Slowly it dawned on the com-pany's board that with Annesley and his men imprisoned in Surat—and Bombay vulnerable to imminent attack—the crisis in India was just as dangerous to the company's future as the calico backlash at home. Orders were sent to company settlements throughout India to detain and interrogate the crews aboard any ships in the region, seeking information on the whereabouts of Every. Anyone detained who had deserted a pirate ship was to be dispatched back to London as a potential informant.

Unfortunately for the company, it was an unusually challenging time to mobilize a response to an overseas crisis. Previously such issues would have been presented to the Lords of Trade, but that administrative body happened to be in the middle of a transformation into a new institution that would be known as the Board of Trade, with a permanent professional class of civil servants. The East India Company had a compelling case that a strong response from the British government was in the interest of both the company and the nation. The problem was that the institutional body responsible for hearing that case was still in the process of self-assembling.

Eventually, though, the East India Company recognized that its own resources were insufficient to conduct a dragnet of such vast scale. On June 19, twenty of the company's directors gathered at East India House on Leadenhall Street for what was then called a "court of committees"—the equivalent of a corporate board meeting. The attendees included some of London's most powerful merchants and political figures, among them George Bohun, governor of the East India Company and member of Parliament, and Sir John Fleet, former Lord Mayor of London. A third of the men in the room had been knighted; most of them had accumulated significant fortunes from the rise of company stock over the preceding decade and stood to suffer meaningful losses if the Every crisis resulted in the expulsion of the company from India.

Also at the table in the committee room was a company director just shy of his fifty-eighth birthday named Isaac Houblon, the brother of James Houblon, whose catastrophic Spanish Expedition Shipping had initially provoked Every's mutinous turn to piracy. Houblon joined the meeting with a double motive: to protect the East India Company's interests in India, and to seek some sort of restitution for his brother's financial losses—starting with the theft of his prize vessel, the *Charles II*.

The committee discussed a few routine matters—a minor dispute over customs, the payment of a large bill to a trader—before turning to the main topic of conversation. The court minutes make note of the "English pyrate" Henry Every and his "great depredations on some of the ships of the great Mughall in the Red Sea which may greatly prejudice the company's affairs in those parts." The directors agreed to form a special committee with the mandate "to advise what means are most proper to be used for apprehending said pyrate, whether by letter of marque and commission under the Great Seal of England, or by his Majesty's Proclamation, or what other method may be prescribed in the case for vindicating the company's honor and innocence in this matter—and for declaring their abhorrence of such detestable practices." The committee was also instructed to write a letter expressing that innocence and abhorrence to Aurangzeb himself. Four directors were named to the committee; one of them was Isaac Houblon.

Within weeks, the company's secretary, Robert Blackborne, had submitted a handwritten petition to "Their Excellencies The Lords Justices of England." The document reviewed Every's crimes, quoted at length from the pirate's 1695 letter, and described the imprisonment of Annesley and the other company factors in Surat. It warned of "many great inconveniences which may be made upon them at Surat or other factories from reprisals which may be made upon them at Surat . . . But also from the interruption which may be there given to their trade from port to port in India as well as their trade to and from thence to England." It ended with a simple plea: "We do must humbly beseech your Excellencies to use such effectual means for the preventing [of] the great loss and damage which threaten them hereby as your Excellencies' great wisdom shall be thought fit."

The petition generated a quick response from the Lords Justices, who immediately issued a formal proclamation, the opening lines of which established a clear demarcation between the laws of England and the depraved actions of the pirates:

> Whereas we have received information from the Governor and Company of Merchants of London Trading to the East Indies that one Henry Every, Commander of the Ship called Fancy of forty-five guns and one Hundred and thirty men has under English Colours acted as a Common Pirate and Robber upon the High Seas and hath performed under such Colours to commit several Acts of piracy upon the seas of India and Persia and may occasion great Damage to the Merchants of England . . . We do here by charge command all his Majesty's Admirals, Captains and other officers at sea, and all his Majesty's Governors and Commanders of any forts, castles or other places in his Majesty's planation, or otherwise, to seize and take Henry Every and such as are with him on the ship to be punished as pirates upon the High Seas.

A subsequent proclamation, composed by the newly formed Board of Trade, based on the original draft from the Lords Justices and signed by King William himself, extended the vast collection of authorities—and indeed, ordinary citizens—encouraged to participate in the hunt for Every and his men:

> We do therefore, with Advice of the Lords of Our Privy Council, Require, and Command, the Sheriffs of the several

Shires, Stewarts of Stewartries, Baillies of Regalities, and their Respective Deputs, Magistrats of Burghs, Officers of Our Army, Commanders of Our Forces and Garisons, and all others Employed, or Trusted by Us in any Station whatsoever, Civil or Military within this Kingdom, and Our Good Subjects whatsoever within the same, to do their outmost Indeavour and Diligence to Seize upon, and Apprehend the Persons of the said Henry Every, and several of his Accomplices . . .

After further back-channel negotiations with the special committee on piracy, the government added a reward worth more than $50,000 in today's currency, quietly funded by the East India Company itself.

The proclamation also offered a smaller reward for the capture of any of the crew of the *Fancy*. The combination of the reward money and the direct address to "any other of Our Good Subjects" sent a not-so-subtle signal to the wider community at sea. Pirates themselves were as welcome to the bounty on Henry Every's head as the sheriffs and commanders were. And if they had to physically harm Every or his men in attempting to capture them, they were welcome to do that as well; the document went on to "indemnify hereby all and every one of Our Subjects from any Hazard of Slaughter, Mutilation, or other Acts of Violence which they may Commit against the said Henry Every." If it were necessary to kill the villain in cold blood in order to apprehend him, the bounty hunters had the permission of the British crown to pull the trigger.

For the first time in history, a global network of military forces, local law enforcement officers, governors in remote colonial outposts, and merchant shipmen—along with amateur bounty hunters, many of them pirates—would be on the lookout for a single wanted man. But if this global manhunt was a preview of coming attractions—the forerunner of the hunt for modern "enemies of all mankind" like Osama bin Laden—it was nonetheless handicapped by the sluggish communication channels of its day. The relay time of news traveling by ship from Bombay to London and back, and the bureaucratic delays introduced by the formation of the Board of Trade severely limited the government's responsiveness to the crisis. Despite all the renunciations of Every's "horrid villainy" and the enticements of the reward money, from Henry Every's perspective, the most important line in the original proclamation came at the very end: the date of its signing, July 17, 1696.

By the time the authorities finally managed to put a price tag on his head and launched the global manhunt in earnest, Henry Every had enjoyed a ten-month head start.

Part Four

THE
CHASE

23

THE GETAWAY

The Indian and Atlantic Oceans
Fall–Winter 1695

fter we had done as much as we thought convenient," John Dann would later chillingly say of the plundering of the *Gunsway*, "we sent her to Surat with the people in her." Watching the *Gunsway* limping toward the mainland, Henry Every would have had one thought dominating his mind: after fifteen months patiently waiting for the perfect opportunity, he now faced a countdown clock. The *Fath* had likely already made its way back to Surat; the *Gunsway* would arrive in a matter of days. All of the Grand Mughal's operatives—not to mention the East Indiamen—would be on the lookout for him by the time the last of the passengers made their way off the treasure ship. Every and his crew had to escape the scene of the crime as fast as the *Fancy* would carry them.

But first, they would have to divide up the loot. The three remaining ships—the *Fancy*, the *Pearl*, and the *Susanna*—had possessed a strength in numbers fighting the Muslim treasure fleet, but now three ships clustered together was a liability, particularly for Every, who knew he could leave the others behind the horizon line within a half day of good sailing, if the winds were favorable. But he couldn't leave them behind before distributing the fortune they had extracted from the *Fath* and the *Gunsway*.

If the pirate articles were clear about anything, it was that the act of dividing up the booty was as close to a sacrament as anything else in the code. It had to be done fairly, or the whole enterprise of piracy would lose its financial allure. Immense risk, unbearable living conditions, and a very real chance that you would end your life disemboweled in the middle of an ocean, five thousand miles from home—you could tolerate all those horrors knowing two things: first, that you were part of an enterprise that could potentially produce immense profits in a matter of months; and second, that the organization would distribute that fortune equitably among all the participants. The pirates were men living in a world dominated by people like Lord Houblon or Aurangzeb, heirs to dynastic wealth that went back many generations: the number of commoners who had escaped their roots and made their own fortune, as Every had declared to Captain Gibson fifteen months before, was vanishingly small. That was the great promise of the pirate's life: You could break free from the cycle of servitude and poverty. But you could only do it if the spoils were equitably shared.

Dividing the spoils was no easy task. There were likely a dozen different currencies collected from the hulls of the two ships; distributing them fairly among the crew would be close to guesswork, without any expertise in or access to current exchange rates. And the rest of the loot would have been even harder to value: jewels, ivory tusks, silk, spices. Quartermaster Dawson would need a few days to get the ratios right, and the act of distributing the booty would itself take time, with the three ships at their most vulnerable: anchored together, their men distracted by the gleam of gold and silver.

Every directed the three ships to sail south, still following the

Indian coast. Near Rajapur, a trade center south of Bombay where there was a small outpost of the East India Company, they took on water and provisions, while Dawson doled out the goods. The ultimate distribution, across the three ships, had multiple tiers: "Some [got] a thousand pound, some six hundred, some five hundred, and some less, according as the Company thought they deserved," Philip Middleton later reported. As one of the youngest members of the crew, Middleton was awarded "above a hundred pound," though he claimed his shipmate John Sparkes later "robb'd him of it."

For an ordinary sailor, £500 would have been close to a lifetime's worth of wages. Recall that most members of the Spanish Expedition, an unusually well-funded enterprise, had been promised something in the order of £3 per month. At that rate, £500 would be the equivalent of ten years of uninterrupted work for a top-shelf enterprise. You could work thirty years in the Royal Navy for the same wages. While the folklore version of Henry Every's story would later claim that the pirate acquired enough gold in the *Gunsway* attack to live as a "pirate King" for the rest of his life, his official share from the heist was £2,000. It was enough money to secure a life of leisure for the remainder of his days, but still short of dynastic wealth.

Of course, to enjoy that life of leisure, Every would have to slip free of the dragnet that was surely coming for him. Every and his men obviously couldn't sail the *Fancy* back to the Thames dockyards and stroll off the ship as heroes, the way Drake had done. To escape the law—and the bounty hunters who might also be on the lookout for them—they would need to get rid of the ship and somehow launder the money they had pilfered from the treasure fleet. The three ships first sailed southwest across the Indian Ocean to the island of Réunion, then known as Île Bourbon.

While nearby Madagascar had been settled by humans more than a millennium before, somehow the mountainous terrain of Réunion had repelled would-be settlers. When the Europeans first discovered it in the 1500s, the entire island was uninhabited, making it one of the very last islands to be settled by human beings anywhere in the world. The French had established a permanent base there in the mid-1600s, and would eventually develop a long-standing and profitable industry on the island growing vanilla, which was, for a time, one of the most valuable commodities on the planet. To help develop the rough terrain and build their plantations, the French captured or bought slaves from Madagascar. By the time Every and his crew arrived on the island in the fall of 1695, the island had developed a reputation comparable to Madagascar's as a pirate's nest, and as a hub for the growing slave trade.

Anchored in Réunion, Every presented himself to the French as a slave trade interloper, the line of work that he had pursued in the years before he signed up for Spanish Expedition Shipping. (Interlopers were themselves borderline criminals, but it was better to be seen as a run-of-the-mill interloper than the world's most wanted pirate.) Reverting to his Benjamin Bridgeman alias, Every used some of the *Gunsway* treasure to purchase ninety slaves. Most of the men Every shackled in the hold would have been born on Madagascar and only recently brought to Réunion by the French settlers. It is sobering to imagine the dislocation and horror of such an itinerary: Born in an island village where your ancestors have lived for centuries, captured by the French and transported to a rugged and remote island that had rejected all human settlement for its entire existence, compelled to work transforming its volcanic soils into some kind of functioning agriculture. And then one day you wake up to find yourself sold to a British pirate, locked up belowdecks on a ship bound for an unknown destination.

In purchasing the ninety slaves on Réunion, Every was performing a crucial job familiar to most master criminals: money laundering. As horrific as it now sounds, slaves were the closest thing to universal currency in the trading centers they were likely to visit, and unlike the "Arabian gold" in the pockets of the *Fancy*'s crew, the slaves betrayed no connection to the *Gunsway* heist. If Every was going to adopt the guise of slave trade interloper Benjamin Bridgeman, a hundred Africans manacled in the hold made the story that much more plausible.

Every also needed the extra labor the slaves provided on board because almost fifty of his men chose to disembark on the island for good, perhaps planning to make their way to the celebrated pirate utopia of Madagascar. Half of them were French and a third were Danes. "They were afraid, if they came to England, and were caught they would should be hanged, and they thought themselves there secure," Middleton later explained. But Every had another itinerary in mind. He had firsthand experience with the corrupt colonial outpost in the Bahamas; if they could make the five-thousand-mile voyage to New Providence (now Nassau) without detection, they might be able to ditch the *Fancy* and then disperse. According to John Dann, some of the crew opposed Every's plan and threatened to mutiny, arguing that they would be safer in the French settlement of Cayenne in South America than in New Providence among the British, however corrupt they might be. "But Captain [Every] withstood it," Dann noted. The *Fancy* would sail to the Bahamas.

A conventional voyage from East Africa to the West Indies would be broken up with multiple stops at friendly harbors along the way—the Dutch Cape Colony on the southern tip of Africa, or the British outpost at St. Helena Island in the South Atlantic—to take on water and other provisions. But Every had to assume that a bounty was already on his head. (Unbeknownst to him, the Royal Proclamation was still six months away.) They would have to make it to the Bahamas somehow avoiding all contact with European ports.

Parting ways with the *Pearl* and *Susanna*, and at last liberated to sail at full capacity, the *Fancy* rounded the cape and made its way to the tiny, uninhabited island of Ascension, a thousand miles west of Africa. (While much of the lore of Every's pirate career revolves around the speed of the *Fancy*, the navigational prowess evident in his journey should not be underestimated.) Their arrival at Ascension happened to coincide with the nesting season of the island's giant sea turtles; the crew brought fifty of them on board and lived almost exclusively on turtle meat for the remainder of their voyage. Amazingly, seventeen men opted to remain on Ascension, preferring life as virtual castaways on one of the planet's most remote islands to risking arrest by the British authorities in New Providence.

In the final days of April, the *Fancy* reached the outer islands of the Bahamas, no more than a day or two from New Providence itself, with 113 free men and 90 slaves on board. Every's risky strategy of avoiding all the traditional layover sites for provisioning had paid off, though only barely. As Every contemplated his final approach to the New Providence harbor, only two days of rations remained in the galley.

24

MANIFEST REBELLION

Bombay Castle

Late 1695

After Khafi Khan had completed his interviews with the *Gunsway* survivors at Surat, he continued on his original mission, porting goods for Abdur Razzak, the commander of Rahiri. Khan followed the coastal route south from Surat, and sometime in the late fall of 1695, he found himself on the outskirts of Bombay. Razzak turned out to have an old connection with John Gayer, and he had taken the liberty to write the East India Company governor to alert him that his emissary would be in the region. Perhaps the two men could meet and attempt to reach some kind of resolution to the current standoff, Razzak suggested. Barricaded in Bombay Castle, waiting for Aurangzeb to launch his inevitable attack on the company headquarters, Gayer leapt at the opportunity to make his case directly to Khan. He sent word to the brother of his chief of staff, who delivered an invitation to Khan in person, proposing a summit between the two men on the grounds of Bombay Castle.

Reading Khan's epic history of the Aurangzeb era now, his antipathy toward the British traders jumps off the page. ("During these troubles," he writes of his entente with Gayer, "I, the writer of this work, had the misfortune of seeing the English of Bombay.") But the contempt he felt for the company did not compromise his characteristically perceptive reporting skills. The account he left behind of his visit with Gayer gives us an unparalleled glimpse of the negotiations between the British and Mughals at the very height of the crisis.

Khan's first sight, on entering the Castle grounds, was an imposing display of the company's military guards standing at attention in full dress:

> Every step I advanced, young men with sprouting beards, handsome and well clothed, with fine muskets in their hands, were visible on every side. As I went onwards, I found Englishmen standing, with long beards, of similar age, with the same accouterments and dress. After that I saw musketeers, young men well dressed and arranged, drawn up in ranks. Further on, I saw Englishmen with white beards, clothed in brocade, with muskets on their shoulders, drawn up in two ranks, and in perfect array. Next I saw some English children, handsome and wearing pearls on the borders of their hats.

In all, Khan estimated that he passed seven thousand musketeers, a number that seems high given the scale of the British operation at that moment in history. After passing through the gauntlet, Khan was ushered directly to Gayer's offices, where the governor greeted him with an embrace and offered him a chair. (Presumably they had a translator between them, but Khan makes no mention of it.) The two men chatted for a few minutes about their mutual acquaintance Abdur Razzak; Gayer tried to establish a civil tone by singing Razzak's praises to his emissary. Before long, though, the conversation turned to the more urgent—and contentious—issues of the day. Why, Gayer asked his guest, were his factors in Surat still in irons?

Khan replied, poetically, "Although you do not acknowledge that shameful action, worthy of the reprobation of all sensible men, which was perpetrated by your wicked men, this question you have put to me is as if a wise man should ask where the sun is when all the world is filled with its rays."

Gayer pushed back. "Those who have an ill-feeling against me cast upon me the game for the fault of others," he said. "How do you know that this deed was the work of my men? By what satisfactory proof will you establish this?"

Here Khan was on firm ground, at least in terms of his access to the participants. "In that ship I had a number of wealthy acquaintances, and two or three poor ones, destitute of all worldly wealth," he explained. "I heard from them that when the ship was plundered, and they were taken prisons, some men, in the dress and with the looks of Englishmen, and on whose hands and bodies there were marks, wounds and scars, said in their own language, 'We got these scars at the time of the siege of Sidi Yakut, but today the scars have been removed from our hearts.' A person who was with them knew Hindi and Persian, and he translated their words to my friends."

Gayer laughed openly at the accusation, but he did not deny the facts as Khan presented them. "It is true they may have said so," he conceded. "They are a party of Englishmen, who, having received wounds in the siege of Yakut Khan, were taken prisoners by him." But they were not employees of the East India Company, Gayer explained, and the company itself had repudiated their actions in the strongest terms.

Initially, Khafi Khan countered with flattery, smiling at Gayer and saying, "What I have heard about your readiness of reply and your wisdom, I have [now] seen. All praise to your ability for giving off-hand, and without consideration, such an exculpatory and sensible answer!" But then, beneath the smile, Khan delivered a threat, referencing the fact that the company had printed coins with the English king's face on them: "But you must recall to mind that the hereditary Kings of Bijapur and Haidarabad and the good-for-nothing Sambha have not escaped the hands of King Aurangzeb. Is the island of Bombay a sure refuge? What a manifest declaration of rebellion you have shown in coining rupees!"

Here as well, Gayer chose not to dispute the facts of Khan's brief. Instead, he turned the tables by casting blame on the Hindustan currency. "We have to send every year a large sum of money, the profits of our commerce, to our country, and the coins of the King of Hindustan are taken at a loss," he explained. "Besides the coins of Hindustan are of short weight, and much debased; and in this island, in the course of buying and selling them, great disputes arise. Consequently we have placed our own names on the counts, and have made them current in our own jurisdiction." The British had nothing against Aurangzeb as a sovereign, Gayer argued. They just needed a stable currency, as businessmen.

The conversation appears to have ended in a stalemate. But at least the two sides were talking. Gayer may have hoped for a more extended visit, but Khan decided it was best to keep the exchange professional: "[The Englishman] offered me entertainment in their fashion," he wrote, "but I accepted only air . . . and was glad to escape."

25

SUPPOSITION IS
NOT PROOF

Nassau, The Bahamas
April 1, 1696

For the generation of pirates that came to prominence in the first decades of the eighteenth century, the town of Nassau would serve as both safe harbor and a tropical pleasure dome of unchecked debauchery. But in 1696, the capital of New Providence island was a village fighting for survival. Originally called Charles Town, the village had been burned to the ground by the Spanish in 1684. The year before Every arrived, the Bahamas' proprietor governor Nicholas Trott renamed the settlement Nassau, after William III's original title, Prince of Orange-Nassau. Trott's rebuilding efforts had been hindered by the long war with the French, which had severely reduced the influx of trade to the islands. The village was so starved of resources that it had not yet been able to construct a proper pier in its harbor. The French had just taken the nearby island of Exuma; rumors circulated that the Bahamas were next in line. Trott did have at his disposal a newly constructed fort, with twenty-eight guns. But Trott had no warships to fend off a French naval assault, and with only sixty residents on the island, he barely had enough men to operate the cannons in the fort.

Those bleak circumstances must have weighed heavily on the proprietor governor's mind when a mysterious longboat rowed into the harbor on the first day of April in 1696, bearing an unusual offer from a slave trader named Benjamin Bridgeman.

Henry Every had enjoyed the luxury of many months to ruminate on what his eventual strategy would be once the *Fancy* finally reached the Bahamas. He had been entirely isolated from any form of human communication, beyond the two hundred or so men on board the ship, for almost the entire year. With no access to news reports—or even the word-of-mouth gossip exchanged while provisioning at friendly harbors—Every had no way of determining his status as a fugitive from justice. Seven months had passed since the *Gunsway* attack. Perhaps the outrage had subsided back in India; perhaps the authorities in Nassau had heard nothing of his exploits. Even before its heyday as a pirate's den, Nassau had had a reputation as a town that operated at the blurred edges of British law, usually turning a blind eye to pirates or slave interlopers. It was even possible that Every and his men would be greeted as heroes. But it was equally possible that they would be greeted as wanted men, enemies of all mankind.

With his characteristic prudence, Every decided to test the waters first. He anchored the *Fancy* north of the deserted Hog Island, out of sight of the New Providence harbor. (Hog Island was rebranded in the 1960s as Paradise Island, and now houses the sprawling Atlantis vacation resort.) Every called the men on deck and outlined his plan. The crew of the *Fancy* would attempt to buy the protection of the Nassau governor by offering him a bribe. All the sailors would contribute a portion of their holdings to the fund: twenty pieces of eight and two pieces of gold. Respecting the pirate's tradition of equitable profit sharing, Captain Every contributed twice what the other men donated to Trott's payoff. As always, whether the pirates were plundering or bribing, the articles of agreement were sacrosanct.

Every wrote a letter to Trott, adopting his old slave-trading persona. (The fact that he was carrying ninety slaves on board the *Fancy* made the alias even more persuasive.) Philip Middleton later claimed to have read the letter with his own eyes. The terms proposed to Trott involved a simple quid pro quo: "Provided he would give them liberty to come on shore and depart when they pleased," Middleton recalled, "they promised to give the said Governor twenty pieces of eight and two pieces of gold a man, and [the *Fancy*] and all that was in her."

The *Fancy* had served Every and his men spectacularly over the nearly two years that had passed since the mutiny at A Coruña. Now she was a liability. She had once been so snug that Every "feared not who would follow her." But he had different kinds of fears now. He no longer needed to outrun his enemies. Now it was time to disperse.

One of Every's top officers, Henry Adams, boarded the longboat with a handful of other sailors, and bearing the note from Benjamin Bridgeman, they rowed their way into the New Providence harbor. The proposal must have sounded suspicious when Trott first laid eyes on it. Why would a British captain give up so much—including his ship—simply for the opportunity to enter the harbor? Trott must have assumed that the gold and pieces of eight being offered to him in tribute had not been acquired through legal means. The *Fancy*—"and all that was in her"—were undoubtably stolen goods. In Trott's defense, however, the proclamation announcing the global manhunt for Every would not be issued for another three months. Situated eight thousand miles from Surat, with the usual communication channels with Europe compromised by the French Navy, it is

entirely possible that Trott had heard nothing of the *Gunsway* contro-
versy. He must have known that agreeing to Benjamin Bridgeman's
terms would entail going into business with a pirate. But he most
likely had no idea that the pirate in question happened to be the most
notorious one on the planet.

Bridgeman's offer had an appeal that went beyond the economic
reward of the bribe itself. He had no way of knowing, from reading
Bridgeman's letter, just how "snugg" the *Fancy* was, but having a
warship with forty-six guns to defend his fledgling colony would
give him a significant new resource if the French did attack the island. And the influx of men would, overnight, triple the town's population. Even if only a fraction of Bridgeman's crew remained at
Nassau, Trott might well have enough manpower, with a warship
and the new fort, to put up a legitimate fight against the French.
Surely the authorities back home would prefer that he negotiate a
deal with an Englishman—however shady his past—if it meant holding on to a promising new outpost in the West Indies.

Whatever his moral calculations, Trott sent a response back,
writing in "very civil terms," Middleton later reported, "assuring
Captain Every that he and his company should be welcome." They
had a deal: the crew of the *Fancy* could come ashore at their liberty,
and in return, Trott would get a warship, forty-six cannon, and a
small fortune in stolen goods.

More than a year later, when it became apparent that the governor had not only given sanctuary to the world's most wanted man
but also taken possession of a stolen ship, the original investors in the
Spanish Expedition sued Trott, attempting to recover in damages
some of the losses from their catastrophic venture. In his deposition,
Trott claimed that the town had no choice but to welcome Every to
the island. "Even if the people of Providence had been stronger," he
testified under oath, "it would have been necessary to have invited

the ship in, for on the 4th of April the French had taken the [nearest] of the salt ponds and were meditating an attack on Providence, had they not heard of the arrival of this ship, which had 46 guns." Asked whether he realized at the time that the men were pirates, Trott professed his innocence: "How could I know? Supposition is not proof."

"Not long thereafter, a great ship rounded Hog Island," the historian Colin Woodard writes, "her decks crowded with sailors, her sides pierced with gun ports, and her hull sunk low in the water under the weight of her cargo. Adams and his party were the first to come ashore, their longboat filled with bags and chests. The promised loot was there: a fortune in silver pieces of eight and golden coins minted in Arabia and beyond." According to Middleton, the crew of the *Fancy* left behind "fifty Tons of Elephants [tusks], forty six Guns mounted, one hundred Barrells of Gunpowder or thereabouts, [and] severall Chests of Buccanneer Guns."

As the crew finished unloading their goods, Henry Every rowed ashore in the longboat and was greeted by Nicholas Trott. The two men then retired for a private conversation.

Henry Every would not set foot on the decks of the *Fancy* again.

26

THE SALTWATER
FAUJDAR

Surat, India
Winter–Spring 1696

S hackled "like a company of Dogges" in the Surat factory, Samuel Annesley had been stripped of his personal liberty, but he did have one luxury at his disposal: a near infinite supply of time to think. As the imprisonment stretched into the winter of 1695, an idea began to form in his mind. Perhaps the company could turn the catastrophe of the *Gunsway* attack into an opportunity. The idea might have been originally triggered by the formal language that Aurangzeb had issued, once Khan's account of the raid reached him: if the East India Company wanted to continue its business on the subcontinent, it would have to "find out the pyrates that infest the seas, or else to make satisfaction with the merchants for their effects robbed from the *Gunsway* for which to give security to that value and with the merchant ships a cannon to accompany them, as no mischief comes to them whether they are bound, if any damage ensues to make satisfaction."

The company was already committed to "find out the pyrates that infest the seas," or at least to finding one pirate in particular. The key phrase in the Grand Mughal's edict came in the second half, where Aurangzeb proposed that the company "give security" and "a cannon to accompany" the merchant ships of Muslim India. Aurangzeb presented it as a concession that the British would have to make in order to stay in his good graces. But Annesley saw it as something else: an opening. In a letter to Gayer, Annesley drew an analogy to the Grand Mughal's *faujdars*, the military officers assigned to serve as law enforcement in specific territories across the land. The East Indiamen were already among the most imposing vessels in the water, ideally suited to protect the ships of Aurangzeb and merchants like Abdul Ghaffar. Perhaps the Grand Mughal could be persuaded to grant the company the same authority at sea that he granted his police commanders on land? "As his land *faujdars* are responsible on account of their salaries to make good robberies on the roads," Annesley proposed to Gayer in a letter in late 1695, "[the company could] satisfy all such losses on the salt water. These great and noble benefits will sufficiently atone for our present disgrace and loss and set the English nation for ever out of danger of such abuses again." Annesley thought the new role would also help the company's bottom line, suggesting that they might be able to charge Aurangzeb as much as 400,000 rupees per year for their services as a nautical security force.

At first, Gayer resisted Annesley's scheme, in large part because the company, in its weakened state, lacked the resources to provide sufficient protection to Aurangzeb's ships. But Annesley continued to make the case. In his mind, protecting the Mughal convoys was not merely a short-term solution to the *Gunsway* crisis, or a new line of business for the company. He began to imagine an arrangement where the company itself would hold a kind of sovereign power at sea. The ships, he argued, would be "in the Nature of Castles and those under our Command as in our harbours [which] the Lawes of Nature and hospitality obliges us to defend." Deputize the British as the legal authority empowered to protect the rule of law on the Indian Ocean, and the whole balance of power in the region would settle on a new equilibrium. In its current situation, Annesley recognized, the company's authority was restricted to its fragile hold on Bombay island. The shackles and the prison guards would have been a daily reminder to Annesley that even on the grounds of the company's factory in Surat its dominion was severely limited. But if Aurangzeb recognized their authority to protect the Muslim treasure ships and other commercial vessels from piracy, their sovereign power would expand meaningfully, over water if not yet over land.

The historian Philip Stern has argued, persuasively, that this strategy—first dreamed up by Samuel Annesley under house arrest in the Surat factory—would prove to be a critical turning point in the relationship between India and England, one that is inevitably neglected in the traditional version of the rise of the British empire in the subcontinent. The standard account, according to Stern, imagines the East India Company as "a commercial body that only 'turned' sovereign—accidentally, haphazardly, and unwillingly—with the Company's great territorial acquisitions in Bengal following Robert Clive's victory at the Battle of Plassey in 1757 and the assumption eight years later of revenue and governance responsibilities in the Mughal office of diwan." In Stern's retelling of the story, the assumption of sovereign power has much deeper roots, extending back to Annesley's schemes of a "salt-water *faujdar.*" "To fight piracy was to claim to be able to draw fundamental distinctions between just and unjust violence, public and private right, and honorable and dishonorable behavior at sea," Stern writes. "It was also to assert the right to enforce those distinctions and exercise a certain form of imperium over the sea lanes."

Gayer eventually saw the merits of Annesley's strategy, and secured the blessing from the court of committee back at East India House to propose a settlement with Aurangzeb: the company would compensate the Grand Mughal for his losses, and assume the responsibility of protecting his ships. In the first weeks of 1696, it seemed as though the prisoners at the Surat factory were on the edge of being released, but the negotiations with Aurangzeb's court dragged on for months, in part because the Grand Mughal himself was still simmering over Every's sacrilegious acts. According to Khafi Khan, Aurangzeb's righteous fury was ultimately sabotaged by I'timad Khan, the Surat governor in Annesley's pocket. "He saw all these preparations," Khan wrote, "and he came to the conclusion that there was no remedy, and that a struggle with the English would result only in a heavy loss to the customs revenue." Annesley's biographer, on the other hand, credits the Indian politician Asad Khan, who recognized that the "renewal of warfare against the [company] would probably be a very onerous business, and that in any event it would react disastrously on the Imperial revenue." Both men appear to have ultimately come around to the argument Gayer had been making from the outset: that the company was too much of a profit center for the Mughal regime to expel it. "After sundry false alarms," according to Wright, "the welcome tidings of the arrival at Surat of the court's orders for the reopening of the port reached the factory on June 27 and the same day the Governor caused the irons to be removed from the prisoners and the guard to be withdrawn." Nine months after Henry Every's assault on the *Gunsway*, the East India Company was back in business.

During the months leading up to that release, as John Gayer weighed his options in Bombay Castle and Annesley tugged at his chains in Surat, it would have seemed beyond fantasy that the East India Company would become an imperial force on the subcontinent barely sixty years later: a company/state ruling over a hundred million subjects. But Annesley's vision and Gayer's negotiating skills transformed the crisis into an opportunity for the company to expand its dominion. What seemed like an existential crisis for the East India Company became, against all odds, the first stirrings of empire.

Alternate histories are, almost by definition, indistinguishable from fiction, but here the narrative requires only a few small tweaks to result in a radically different outcome. Had a deal not been struck with Aurangzeb to resolve the *Gunsway* crisis, had Samuel Annesley not perceived the opportunity lurking in the Grand Mughal's initial demands, the East India Company might well have been forced to abandon Surat and Bombay. Stripped of its cash cow, and vilified by the weavers and other protectionist groups, the company could easily have collapsed under the weight of challenges at home and abroad. No doubt a loose network of successor firms would have been back at Surat and Bombay within a matter of years, rubbing shoulders with the Dutch and the Portuguese. But would the English have ultimately conquered India if dozens of smaller firms had served as the commercial conduit between the two nations? It is impossible to say for sure, but certainly the odds would have been worse.

The English had been traders in a strange land—sometimes welcome, sometimes on the edge of exile—for more than eighty years, ever since Jahangir had first granted them the ability to "sell, buy, and to transport into [the] country at their pleasure." But now, watching over the Muslim treasure fleet as it made its pilgrimage to Mecca, keeping the seas free of pirates, the English had an asset they had not possessed before, a new power that would come to define their relationship to the subcontinent: the force of law.

27

HOMECOMINGS

Dunfanaghy, Ireland

Late June 1696

The men were never going to stay long in Nassau. They were big fish now, too big for a backwater outpost with sixty residents. As they spent a small slice of their fortune in the two taverns in Trott's struggling village, they would have known they were vulnerable there, clustered together under the good graces of a corrupt British governor, with a dragnet stretched across the planet seeking them out. Nicholas Trott did his best to persuade the men to stay, hosting a feast for the prospective Bahamians at his home. ("One of the men broke a drinking glass," Philip Middleton later recalled, "and was made to pay eight checqueenes for it.") Trott's respect for the new arrivals, however, did not extended to the *Fancy* itself. After Every's gang had handed the ship over to the governor, he assigned responsibility for the ship to men whose "incapacity or number were not sufficient to secure her from hurtfull accidents," as a subsequent missive to the Board of Trade reported. James Houblon's "extraordinary sailor" ended her short but eventful existence as a wreck, never to sail again. Phillip Middleton would later call it a "sad sight"—the ship that had outrun all its enemies, that had kept the men alive on a journey of more than ten thousand miles, now foundered in the shallow waters of the Nassau harbor.

Nicholas Trott's hospitality persuaded six or seven of the Every gang to stay, to disappear into the small-town life of a remote colonial outpost and be forgotten. There appears to have been some sexual intrigue shaping the decision as well. A few of the men who remained behind married Nassau women. (Quartermaster Adams seems to have married one of them in a matter of days.) But the rest were eager to move on.

Within a week or two, the men had sorted into three distinct escape pods. One group of twenty-three pirates acquired a sloop in town; they embarked on a return trip straight back to England, assuming they could slip through the disorganized border control on the Thames docks and slink their way back to their families and loved ones. The largest group opted for the same strategy of seeking out pirate's nests that had steered them to Madagascar and Nassau—only this time on a grander scale. Those pirates went to the American colonies.

The pull toward the American mainland was partially a simple matter of proximity. Charleston was only four hundred miles away. But there were legal reasons to head to America as well. The colonies had developed a reputation for both nurturing and tolerating piracy, a reputation that Rhode Island's Thomas Tew had amplified just a few years before with his 1693 Red Sea robberies.

The reputation turned out to be a valid one, as least as far as Every's gang was concerned. Not one of the fifty men who sailed for the Carolinas were ever convicted of crimes associated with the *Gunsway* attack. Some had brushes with the law; some disappeared. But not a man among them was ultimately punished for his crimes. According to some accounts, Every's crew openly boasted about their heroic days on the Indian Ocean. In early 1697, James Houblon received a letter from a scandalized colonist in Pennsylvania who had overhead Every veterans "regaling their fellow patrons" at a tavern with the stories of their exploits aboard the *Fancy*. The atmosphere was so lax that the pirates barely bothered to conceal their identities. "They brag of it publicly over their cups," Houblon's correspondent noted.

The colonies had another asset in their favor: a thriving market for slave labor. Presumably some of the slaves that that had been captured in Guinea or acquired in Réunion traveled with the crew to the Carolinas and were sold off along with the remnants of the *Gunsway* treasure. According to Philip Middleton, some of the slaves were sold in Nassau to Trott and his men. Assuming some of them stayed on the island, they would have played an early role in the demographic transformation of the Bahamas, a country where today more than 80 percent of the population is of African descent—one small piece of the vast diaspora that slavery produced.

Two months after making his initial offer to Nicholas Trott, Henry Every left Nassau accompanied by twenty of his original crew, including Henry Adams and John Dann. With the world's most wanted man in their party, they dared not risk a direct return to England. Instead, Every and the others bought a single-masted sailboat called the *Sea Flower* and set sail for northeast Ireland. In what must have been one of the least romantic honeymoons in history, Henry Adams brought his new bride along for the trip.

Sometime in late June, the *Sea Flower* sailed into the small harbor of Dunfanaghy, nestled at the western edge of Sheephaven Bay, roughly a hundred miles northwest of Belfast. According to Dann's account, on their arrival they were confronted by a "landwaiter"—effectively a customs official—who eventually allowed them to continue their travels toward Dublin in exchange for a bribe of £3 per man. Dann traveled with Every—who was still using the alias Benjamin Bridgeman—for six miles, before the captain announced that he was going to break off from the company and head out on his own. "I heard he went over for Donaghedy in Scotland," Dann later recalled. "I heard him say he would go to Exeter when he came into England, being a Plymouth man."

As Every and his crew dispersed across the British Isles, back in London at East India House the special committee on pirates was ramping up its manhunt efforts. The proclamation from the Lords Justices offering a reward for Every's capture was released several weeks after the *Sea Flower* landed in Ireland. The company paid to have a hundred copies of it printed and shipped off to its factories in India. (They added an additional reward of 4,000 rupees for any Indian informants who led them to Every.) By late July, Isaac Houblon and his fellow committee members had learned that Every had landed in Nassau and was rumored to have left with a small crew, headed back to England and Ireland. The company's secretary, Robert Blackborne, immediately dispatched a flurry of letters to local authorities in port towns across the British Isles, requesting that they be on the lookout for Every and his men. "Captain Henry Every now goes by the name of Bridgeman," Blackborne advised. "It will be an admirable service to the Kingdom and as such recommend to your favor to capture any of them that shall come into your parts." Just as Every's crimes were endowing the company with new naval power in the Indian Ocean thanks to Samuel Annesley's scheme, the manhunt back home bound the company's agents into a close partnership with law enforcement authorities. Blackborne, after all, was merely a corporate secretary, transcribing minutes of board meetings and writing letters to company representatives overseas. But with Every on the loose, he had taken on a new responsibility: issuing an all-points bulletin for the nation's most wanted man. The company arranged to have agents on alert, ready to be dispatched to interrogate and bring back to London any suspects that local law enforcement detained. Whatever political tensions and scandals had compromised the relationship between the government and the company, the threat posed by Every and his crimes forced the two institutions into a united front—so united, in fact, that the East India Company took on many tasks that would have traditionally been delegated to the state.

John Dann continued his travels to Dublin, then sailed to Holyhead in Wales. After a brief sojourn in London, he traveled north to his hometown of Rochester, where he booked a room in a local inn. It turned out to be a disastrous homecoming. Dann had traveled more than ten thousand miles, carrying his profits from one of history's largest heists sewn into the lining of his jacket. But during his first day back in Rochester, an inquisitive maid cleaning his room noticed the unusual weight of the coat while folding his clothes. She reported him to the authorities, who found more than a thousand Turkish coins "quilted up in his jacket." The town mayor seized the coins and threw Dann in jail under suspicion of robbery.

Dann's arrest was just the beginning. Over the course of the summer, seven more men from Every's crew were apprehended in Liverpool, Dublin, Newcastle, and in the West Country near Every's birthplace. The special committee spent close to a thousand pounds in reward money and in "gratuities" doled out to officials who had assisted with the dragnet. They even covered the cost of transporting the prisoners back to London, where they could be tried together in the most public forum possible. The East India Company—and its close collaborators in the British government—had been eager to throw the full force of the law against the *Gunsway* pirates ever since John Gayer's letter detailing the atrocities had arrived London in late 1695. Now they had eight of them in custody. At long last, the world—and Aurangzeb most of all—would have a chance to see England's true position on piracy.

Part Five

THE
TRIAL

28

A NATION OF PIRATES

London

September–October, 1696

Several months after Aurangzeb released the prisoners at Surat, as the East India Company was resuming its business affairs in Southeast Asia, John Gayer sent a letter back to London, reflecting on the *Gunsway* crisis and its implications. The British government, he argued, needed to make a strong stand against piracy, at home and abroad. The Drake era—where pirates were often seen as unofficial agents of British interests abroad—might have worked with nations that were declared enemies, or tribal societies like the Spice Islanders wiped out by the Dutch. But with a genuine trading partner like India, the leniency of the Drake era no longer made sense. Gayer argued that the British needed to extend the same authority to the East Indiamen policing the Red Sea that Aurangzeb had granted them in the post-*Gunsway* pact: "If there not be care taken to suppress pyrates in India and to empower your servants there to punish them according to then deserts, without fear of being traduced for what they have done when they return to their native country, it's probable their throats will be all cut in a little time by malefactors and the natives of the country in revenge for their frequent losses." He ended the argument by appealing to their base economic interests. Turn a blind eye to Red Sea pirates, he predicted, and "your Honours' trade in India [will be] wholly lost."

Fighting piracy, however, was not solely the province of Annesley's "salt-water *faujdar*"; with eight members of Every's crew in custody, the Crown could now make a public case against these enemies of all mankind in the strongest terms possible. The romantic myth of swashbuckling Captain Every had been spread by the balladmongers and pamphleteers. The government, however, had a tool at its disposal that the emergent popular press lacked: a criminal trial.

To be sure, the popular press thrived almost parasitically off the drama and lurid details that criminal trials provided. (A large number of the "ballads" sung by the balladmongers were effectively homicide trial transcripts set to music.) But the barkers and protojournalists couldn't control the trial itself; they could only transmit it, albeit with the usual distortions designed to make the story more salacious to their consumers. In a public trial, the government would be given a megaphone—most powerfully in the opening and closing statements of the prosecution—to establish what the historian Douglas R. Burgess Jr. calls a "dominant historical narrative of piracy":

The crown and its Board of Trade had an image crisis. England was perceived as a "nation of pirates" and was now learning (to its chagrin) that this charge was true, at least in the colonies. Of paramount importance, superseding even the enormity of doing justice to Every's men and re-establishing relations with the Mughal, was to use the trial as a means for making the government position crystal clear: pirates were hostis humani generi . . . not merely enemies of England but of the entire world. It placed the Every pirates in the highest echelon of international criminals and committed the English state to regarding their eradication as a first priority.

The criminal trial also offered the state another key resource in establishing that "dominant historical narrative": capital punishment. The spectacle of Every's men dangling on Execution Dock would send an unequivocal message to the world that England had no tolerance for the international criminals behind the *Gunsway* attack.

All these elements gave the English state a commanding platform not just in a court of law, but, just as important, in the broader court of public opinion—what the German sociologist Jürgen Habermas famously called the "public sphere," the emerging realm of coffeehouse debate, pamphleteering, and sidewalk oration that would play such a defining role in the Enlightenment culture of the eighteenth century. But all those tools—the public theater of a criminal trial, the spectacular violence of a state-sponsored execution—were themselves dependent on recent shifts in the legal jurisdiction that governed cases of piracy. Because the crimes committed by pirates were almost always committed far outside the geographic domain of British law, they had historically been treated as civil law cases adjudicated by the Admiralty. Those trials were not open to the public; they granted the defendants extensive legal representation; and, crucially, the Admiralty lacked the authority to issue death sentences. A late-seventeenth-century *criminal* trial under English common law, on the other hand, shifted the entire balance of power toward the state: not only was capital punishment on the table, but the accused were not allowed to have their own counsel. Trained experts delivered the case for the prosecution, while the defense was supported exclusively by the limited legal expertise of the defendants themselves. And criminal trials could be attended by the general public, their twists and turns and dramatic resolutions transformed into a theatrical narrative by the pamphleteers and balladmongers.

Over the course of the 1600s, as it became increasingly clear that the civil law precedent for piracy cases was limiting the government's ability to successfully prosecute pirates, a series of legal reforms were set in motion, creating a special class of piracy-related crimes that would exist in a strange dual state: technically defined as civil law crimes under the jurisdiction of the Admiralty, they would nonetheless be prosecuted in common law courts, thanks to their extreme nature and the threat they posed to the stability of the English nation and to its trade relations. Had Every's gang committed their crimes a century earlier, the trial would have been a far weaker vehicle for establishing a dominant narrative about piracy—and the men themselves would have had no fear of execution. At the same time, the shift to common law did give the pirates one potential advantage: common law trials were decided by a jury. The pirates' innocence or guilt would be determined not by the elder statesmen of the Admiralty who were predisposed to revile piracy in all its forms, but rather by a jury of ordinary citizens, whose attitudes toward piracy were shaped by the balladmongers and pamphleteers more than the venerable legal tradition of *hostis humani generis*.

Then there was the question of witnesses. None of the actual victims of the *Gunsway* attack were available to testify, of course. And a jury of ordinary British citizens circa 1696 would be unlikely to find a Muslim merchant's account of being robbed at sea particularly sympathetic. (As for the sexual violence on board the *Gunsway*, rape trials were infrequent in seventeenth-century England, and effectively nonexistent if the victim happened to be a foreigner.) If the prosecution was going to make a compelling case for executing Every's men, they would need eyewitnesses drawn from the pirate's gang itself. In other words, they needed to compel at least one of the men in custody to flip.

Fortunately for the state's case, they found such a turncoat in the very first man they had managed to capture: John Dann. It is unclear what means the authorities used to extract a confession from the Rochester pirate, but within days of having the Turkish coins discovered in his jacket, Dann had offered a full account of the Every gang's predations. On August 3, Dann delivered a sworn testimony that relayed the entire sweep of the previous two year's events, from the initial mutiny in Spain to the Indian Ocean attacks to the safe harbor provided by Nicholas Trott—all the way to Dann's arrival with Every in Ireland. The very next day, the Lords Justices in Ireland would hear a similar testimony from Philip Middleton. (Subsequent court committee minutes from East India House reveal that the company made several payments to Middleton's mother to compensate him for testifying against his mates.) Thanks to the legal reforms of the preceding decades, the government had a legal platform that allowed them to demonstrate to the world just how abhorrent piracy was to British values. Now, with Dann and Middleton talking, they had something else: evidence.

With the star witnesses in place, one final question remained for the government to resolve: Which crimes in particular should the prisoners be charged with? The list of potential offenses was long. They had committed mutiny against a British captain and stolen a ship of force owned by prominent Londoners. They had committed acts of piracy on board the *Fath Mahmamadi* and the *Gunsway*; they had raped and tortured the men and women aboard those ships; they had stolen from the English and the Danes in the summer of 1694; they had burned a mosque in Maydh.

The government took several months to prepare its line of

attack. In the end, the objective of extinguishing England's reputation as a "nation of pirates" won out. In consultation with the Lords Justices and the Admiralty, the lead prosecutor, Dr. Henry Newton, opted to orient the charges around the offenses committed against the Grand Mughal. It would be a show trial with a global audience, giving the state—and its collaborators at the East India Company—the opportunity to display to the entire world that piracy would be given no quarter by the British government.

The six men put on trial in October 1696 were guilty of many crimes, but the indictment read against them would only mention one: "feloniously and piratically taking, and carrying away, from persons unknown, a certain ship called the *Gunsway*."

29

THE GHOST TRIAL

Old Bailey, London
October 19, 1696

he northern stretch of Old Bailey road, just inside the original boundaries of the City of London, has ties to the justice system that date back almost a thousand years. The Romans had built one of their seven main gates into the city there, and sometime in the twelfth century, that stretch of the wall was reengineered to house a small prison for debtors and felons. Over time, it came to be known as Newgate Prison. A few centuries later, a medieval courthouse rose on the site, allowing easy transfer of accused and convicted criminals between their trials and their jail cells. The courthouse took its name—Old Bailey—from the wall itself: a "bailey" is the outer enclosure of a castle or a fort. The original courthouse burned to the ground in the fire of 1666. Seven years later, a three-story Italianate courthouse replaced it. An etching from 1675 shows the building's most distinctive feature: the main ground-floor courtroom opened on its eastern side to an outdoor area called the Sessions House Yard. The courtroom had been left open to the elements as a hygienic measure; typhus was so rampant in Newgate Prison that it had acquired the nickname "gaol fever," and it was thought that keeping the court-room flushed with fresh air would protect the lawyers and magistrates from contracting the disease. As a public health intervention, the open-air design had little effectiveness. (Typhus is largely transmitted by the bites of fleas and ticks.) But the architecture of the courthouse had a meaningful impact on the legal system's relationship to the public itself: because the courtroom was open to the elements, crowds of spectators and reporters could gather on the street and follow the proceedings of high-profile trials, sometimes heckling and jeering. More than a few juries were swung by the sound of the mob outside in Sessions House Yard.

The crowds gathered early outside Old Bailey on the morning of October 19, eager to catch a glimpse of the notorious Every gang—and even better, to overhear some of their testimony. In the public space in front of the courthouse, the witnesses John Dann and Philip Middleton mingled with onlookers and court personnel, waiting to be admitted into the courtroom. A brick wall crowned by iron spikes separated Dann and Middleton from six of their former shipmates, clustered together in a space known as the bail dock. Most of them had been in prison, awaiting trial, for more than a month.

Standing in the bail dock, the prisoners could hear the bailiff announce the names and titles of the justices who would oversee the case. The names were likely meaningless to the uneducated seamen, but anyone familiar with Britain's legal system at the time would have immediately recognized how formidable the list was. Sir Charles Hedges, Judge of the High Court of Admiralty, would preside over the case, accompanied by Sir John Holt, the Lord Chief Justice of the King's Bench, the branch of the judicial system that dealt with cases that involved the king himself in some fashion. Chief Justices from the Court of Common-Pleas and the Court of the Exchequer—courts that oversaw common law cases involving private property, such as theft—were seated at the bench as well. The most powerful and accomplished justices, drawn from all the major branches of the English judicial system, had gathered at Old Bailey to oversee the pirate trial. Isaac and James Houblon's brother John, former Lord Mayor of London and inaugural governor of the Bank of England, observed the proceedings as well, accompanied by other dignitaries.

Such exceptional legal firepower had been assembled not to ensure a fair trial for Every's men, but rather to ensure a conviction. The Lords Justices presiding over the trial were the same ones who had issued the original proclamation in July denouncing Every as a "common pirate" who had done "great Damage to the Merchants of England." The justices had worked as partners with Isaac Houblon and the special committee of the East India Company to broadcast news of the manhunt for Every to the farthest reaches of Britain's trade networks; they had welcomed the company's £500 reward for Every's capture. (Earlier in 1696, Judge Hedges had summarily dismissed the lawsuit by the unpaid crew of Spanish Expedition Shipping, ruling in favor of James Houblon and the other investors in the doomed venture.) While technically the prosecution of the pirates would be led by Henry Newton, Justices Hedges and Holt would actively participate in the interrogation of Every's men, making no pretense of their disposition in the case. To translate it into a modern context, imagine if the O. J. Simpson case had been tried in front of the United States Supreme Court, with the judges blithely arguing the prosecution's case—and interrogating Simpson himself—from the bench. This was the legal environment that the Every gang found themselves confronting that October morning in Old Bailey.

Judge Hedges began the proceedings by delivering his initial instructions to the grand jury. They were to provide (or withhold) a "billa vera" approving the indictment. While six men were standing in the bail dock, the indictment was read against seven: the middle-aged steward William May; nineteen-year-old John Sparkes, Edward Forsyth, William Bishop, Joseph Dawson, James Lewes—and one Henry Every. (The subsequent court documents appended a curt "not taken" to explain his absence at Old Bailey.) Judge Hedges gave the jury a brief summary of the facts, explained the unusual hybrid legal jurisdiction that applied to piracy, and dispatched the jurors to assess the plausibility of the prosecution's case. According to the court records, the grand jury returned after "a little time" with an indictment. At Hedges's command, guards ushered the six defendants into the courtroom, to stand trial for their crimes against humanity.

The architecture of the courthouse was explicitly designed to channel and concentrate the flow of power within the space. The decor of the room magnified the authority of the state, with heralds and crests testifying to the solemn authority of the Crown. Nautical icons—including three anchors on a tapestry displayed beneath the bench—signaled the presence of the Admiralty. Rebuilt from scratch only a few decades earlier, the Old Bailey courtroom would almost certainly have been the most richly appointed room the defendants had ever set foot in. And their placement in the room only emphasized their status as outsiders. The judges towered above the prisoners from a dais. A mirror hung above them, reflecting the sunlight streaming in from Sessions House Yard directly into their eyes. The explicit purpose of the contrivance was to illuminate the facial expressions of the accused so that the jury could better assess their honesty or contrition. (Sounding boards also amplified their speech.) On the rare sunny day in London, the effect could be blinding. But even in the city's usual gloom, the message was unmistakable: the accused were on a stage, under surveillance, examined by the discerning eye of the state.

Standing at the bar, the prisoners listened as Judge Hedges read the indictment against them. Five of them maintained their innocence. Only Joseph Dawson pled guilty.

After the swearing in of the petty jury, the chief prosecutor, Henry Newton, in powdered wig and ruffled white collar, rose and launched into his opening statements. Echoing the language of the indictment itself, Newton began with the crimes the accused had committed against Aurangzeb: "The prisoners are indicted for piracy, in robbing and plundering the ship Gunsway belonging to the Great Mughal and his subjects, in the Indian Sea to a very great value." He then relayed a short synopsis of Henry Every's career as a pirate: the mutiny in Spain, the "many and great piracies" in the Atlantic and Indian Oceans, leading up to the *Gunsway* attack.

The facts of the case outlined, Newton then turned to the key argument: England would not tolerate piracy. Here he borrowed extensively from the argument that John Gayer had been making from Bombay Castle for the preceding year: if England wanted to profit from its trade with India, if England wanted to engage in reliable commerce with any nation in the world, the country needed to denounce piracy in the strongest way possible. The *Gunsway* heist, Newton explained, was "likely to be the most pernicious in its consequences, especially as to trade, considering the power of the great Mughal, and the natural inclination of the Indians to revenge." But the jury possessed a unique opportunity to remedy those "pernicious consequences," by delivering "that judgment from you their crime deserves." The crime in question, Newton argued, was in its nature far more serious than conventional robberies, because it threatened not just an individual's property rights, but also the growing network of global trade:

> Piracy . . . by so much exceeds Theft or Robbery at land;
> as the interest and concerns of kingdoms and nations are

above those of private families or particular persons. For suffer pirates and the commerce of the world must cease, which this nation has deservedly so great a share in, and reaps such mighty advantage by. And if they shall go away unpunished when it is known whose subjects they are, the consequences may be to involve the nationals concerned in war and blood, to the destruction of the innocent English in those counties, the total loss of the Indian trade, and thereby, the impoverishment of this kingdom.

The last lines of Newton's opening statement might as well have been lifted directly from Gayer's anguished missives to the East India Company directors back in London. If the jury did not perform its sworn duty and bring the Every gang to justice, Newton argued, the consequences would extend far beyond letting a group of guilty men walk free. It would lead to the "impoverishment of this kingdom."

The stakes of the trial thus established, the prosecution called its two key witnesses. John Dann and Philip Middleton were escorted to the witness box, directly opposite the defendants standing at the bar. Just a few months before, all eight men had been toasting to the success of their heist on a tropical island; now they stood face-to-face in the Old Bailey courtroom, antagonists. Whatever solidarity had held their pirate collective together for the past two years had evaporated under the threat of Execution Dock. They were sworn enemies now.

We do not know what transpired in the subsequent hours. We know Dann and Middleton told a story of piracy on the high seas, that they accused their former partners of crimes against humanity. We know that Newton interrogated the defendants; presumably the six men standing at the bar without legal representation made some attempt to defend themselves, drawing upon their limited knowledge of the law. The rest, however, is conjecture. No trial transcripts were ever published. In fact, the state went to great lengths to suppress any record of the trial at all. This blank spot on the map of Henry Every's career—the vanishing act of the piracy trial—was not simply a casualty of unreliable archives. No record of the trial beyond Newton's opening statement exists because at the end of the trial, the jury returned a verdict that sabotaged the dominant narrative of piracy, sending a message to Aurangzeb that confirmed his worst suspicions about the English. William May, John Sparkes, Edward Forsyth, William Bishop, Joseph Dawson, James Lewes—they were all, to a man, declared not guilty.

Accused of crimes against humanity, accused of violating the property and the direct relations of the Grand Mughal of India, the six men were found by the jury of their peers to be innocent of all charges. Even Henry Every—"not taken" but charged with the crimes nonetheless—had been exonerated.

30

WHAT IS CONSENT?

Old Bailey, London

October 31, 1696

W hy did the state's case against the Every gang collapse? It is possible that some legal stratagem backfired, or the accused—despite their limited education—managed to make an unusually stirring defense of their actions, though subsequent events make both these scenarios unlikely. The most convincing explanation for the shocking acquittal is that the Admiralty and the Lords Justices who had framed the case had underestimated the popular, nativist appeal of the Henry Every myth. The state almost certainly mounted a convincing argument that the pirates standing at the bar had stolen from the Great Mughal and the sovereign nation of India. But in the minds of a jury accustomed to heroic tales of Bold Captain Every and other swashbucklers, with little empathy for a foreign emperor and his subjects five thousand miles away, those actions may have simply not constituted a crime—and certainly not one that warranted a sentence of death.

Whatever the explanation, the verdict was a catastrophe for the state. It was one thing for Every and his men to have avoided detection for a year after the *Gunsway* heist. But to have six pirates under arrest, with two witnesses testifying against them, and still let them walk free? The not guilty verdict confirmed all the allegations against the British state—that for all its tough talk about *hostis humani generis*, the state was either tacitly supporting pirates, or incapable of enforcing the laws against them. The admiralty had planned to use the case against the Every gang as a show trial, a statement to the world that would announce the British government's new zero-tolerance policy for piracy. They had even hired a publisher named John Everingham to release the transcripts of the trial allowing interested readers throughout the British empire who couldn't make it to Sessions House Yard to follow along. Needless to say, Everingham never published the transcripts. One London periodical apologized for its lack of coverage: "We had prepared a more ample account of the Tryal of the Pyrates," the editors noted, "but in compliance with the prohibition of Authority have omitted it."

For the five pirates who had pled not guilty, the acquittal must have seemed like a miracle, given the overwhelming legal apparatus that had confronted them in the Old Bailey. And they must have been perplexed when the court guards escorted them back to Newgate Prison after the verdict, rather than setting them free. For two days, the acquitted prisoners languished in their cells, expecting to be liberated at any moment. During those forty-eight hours a frantic series of conversations unfolded between Judges Hedges and Holt, Prosecutor Newton, and other members of the Admiralty. Double jeopardy prevented them from staging the trial again. It was conceivable that law enforcement would eventually arrest other members of the gang—or Every himself—but even so, word of the first trial would invariably make it back to Aurangzeb, threatening the fragile new alliance Gayer had negotiated with the Grand Mughal. If they were going to make the grand statement—to establish the overarching story that England was no longer willing to turn a blind eye to the predations of the pirates—they were going to have to do it with the men waiting to be released from Newgate Prison.

The solution to this quandary arrived through what historian Douglas Burgess calls a "brilliant act of legal legerdemain": If the Grand Mughal of a distant nation had not made for a particularly sympathetic victim, why not recast the role with a victim that would be more appealing to a British jury? The pirates had stolen from Aurangzeb, but they had also stolen from James Houblon and the investors of Spanish Expedition Shipping. Instead of accusing them of robbing the *Gunsway*, what if the state centered its argument on the theft of the *Charles II*? The men had been acquitted of *piracy*, but the state could still charge them with *mutiny*.

On Saturday, October 31, the six original prisoners found themselves back at the bar at the Old Bailey, listening as a new indictment was read against them. As the new jurors were introduced in the court, Judge Holt made no effort to conceal his displeasure with the verdict rendered in the original trial. "If you have returned any of the former jury, then you have not done well," he thundered from the bench, "for that verdict was a dishonor to the Nation."

In his formal opening statements to the grand jury, Judge Hedges took a more subtle tone, craftily linking the mutiny aboard the *Charles* with the general crimes of piracy itself. "Now Piracy is only a sea term for robbery, piracy being a robbery committed within the jurisdiction of the Admiralty," he explained. "If any man be assaulted within that jurisdiction, and his ship or goods violently taken away without a legal authority, this is robbery and piracy." Whether you were stealing a boat in a Spanish harbor or treasure in the Indian Ocean, you were engaged in acts of piracy. This conflation of piracy and mutiny appeared in the main clause of the indictment: the accused had "upon the high and open seas, in a certain place about three leagues from the Groyn, and within the jurisdiction of the Admiralty of England, piratically and feloniously set upon one Charles Gibson . . . the Commander of a certain merchant-ship called, The Charles The Second."

The prisoners listened to the indictment in utter confusion. Hadn't they just been acquitted of all charges against them? Why were they back in Old Bailey, standing before a judge and jury?

The court officer asked them how they pled, starting with Every's original quartermaster, who had pled guilty in the first trial.

"How say'st thou, Joseph Dawson, art thou guilty of this piracy and robbery, or not guilty?"

Baffled by the situation, Dawson replied, "I am ignorant of the proceedings."

"He pleads ignorance," the officer reported. A clerk reminded Dawson that he had only two options.

"Guilty," Dawson said, returning to his original plea.

Edward Forsyth and William May repeated their not guilty pleas, but when the officer turned to young William Bishop, the disorientation in the court was palpable.

"How say'st thou, William Bishop, art thou guilty or not guilty?"

"I desire to hear the whole indictment read again."

"You have heard it just now," one of the justices replied, "and may hear it again if you desire it."

"The former indictment," Bishop clarified.

"No, there is no occasion for that," the justice replied sharply. "This is an indictment for a fact distinct from that."

In the end, the five men pled out the way they had in the first trial. The jury had found them all not guilty of committing piratical acts against Aurangzeb in the initial trial. Now they would have to determine if the men were guilty of doing the same to James Houblon.

The Advocate General for the Admiralty, Thomas Littleton, rose and addressed the jury, with a ringing condemnation of the defendants. "Their wickedness has been as boundless and as merciless as the element upon which their crimes have been committed," he thundered. What's more, their crimes had slandered England's reputation in the eyes of the world; the entire planet, he claimed with only some exaggeration, "has been sensible of their rage and barbarity."

Joseph Gravet, a second mate of the *Charles*, was the first called to the witness box. Gravet relayed the details of the original mutiny, claiming that Every's men had seized him and locked him in his cabin under armed guard. He described how Every had "kindly" given him a coat and waistcoat when Gravet ultimately decided to leave the *Charles* in the longboat. And then he relayed what would become a crucial piece of evidence. As he was boarding the longboat, Gravet claimed, "William May took me by the hand and wished me well home, and bid me remember him to his wife."

"Was there liberty for any of them that would go ashore?" one of the prosecutors inquired.

Gravet nodded, "Captain Gibson told me so, and there were about seventeen that went off."

"Would the boat hold more?"

"Yes, sir."

THOMAS DRUIT, the first mate of the *James*, took the stand next. He narrated his confused reception of the secret code for the mutiny—"the drunken boatswain"—and his failed attempts to compel the mutineers to return to the *James*.

"I went to command them back," Druit told the jury, "and they refused."

After Druit's testimony, the state called to the witness box

David Creagh, the other second mate of the *Charles*, who had declined to run off with Every. Creagh had taken the honorable route in Spain, but had subsequently been involved in "piratical" activity that had landed him in Newgate Prison under a separate indictment. Creagh recounted his conversation with Every at the helm of the *Fancy*, where the captain asked if Creagh would "go with him." And he delivered another blow against William May's protestations of innocence. "As I was going down [to my cabin], I met with William May, the prisoner at the bar," he explained to the court. "'What do you do here?' says he. I made him no answer, but went down to my cabin. And he said: 'God damn you, you deserve to be shot through the head,' and he then held a pistol to my head."

Creagh went on to describe the exchange between Every and Captain Gibson, and Every's command that Gibson and his followers row back to the mainland in the pinnace. "I heard them order the Doctor be secured, but if there was any more would go into the boat they might."

Once again, the prosecution returned to the crucial question of the empty space in the pinnace. "Was there any room for more in the boat," Conniers asked.

"Yes there was," Creagh replied.

"Was there liberty for any more to go?"

"Yes, my Lord."

AFTER CREAGH'S TESTIMONY, the prosecution turned to its two key witnesses, John Dann and Philip Middleton, men who had chosen—or had been compelled—to stay on board the *Fancy* and who could relate, on the record, the full story of Every's crimes against humanity—even if the crimes officially at stake in the trial were limited to the mutiny at A Coruña. Dann's account goes on for several pages in the trial transcript, including a detailed survey of the *Fancy*'s layover in Madagascar and her misadventures at the mouth of the Red Sea. After Dann described the battle with the two Indian treasure ships, Justice Holt intervened from the bench to ask about the distribution of profits from the heist.

"That was a brave prize, was it not, the best you had all the voyage?" Holt asked.

"Yes, my Lord," Dann confirmed.

"Did you all share?"

"Yes, all that were in the ship."

Holt then asked Dann to confirm that each prisoner standing at the bar had received their share of the bounty. Having implicated the defendants in the robbery of the *Gunsway*, the prosecution invited them to ask questions of their former shipmate. William May seized the opportunity to introduce what would become a key plank of his defense: that he had fallen ill shortly before the *Fancy* had made its final approach to the Red Sea, and had been left behind in the Comoro Islands, only to be reunited at a later point—thereby missing the entire *Gunsway* assault.

"My Lord, may I speak for myself?" May queried Holt.

"If you will ask him any questions, you may. You shall be heard again to speak for yourself by and by," Holt explained.

"My Lord, I desire you will ask him, whether he thinks I had any knowledge of the going away of the ship."

Dann declined to answer: "I have no knowledge of that."

The line of questioning provoked a sharp rebuke from Justice Holt. "You were there and you had a share of the prize," he snapped back at May. "You drank a health to the success of your voyage."

May backed down in response to Holt's outburst: "I hope, my Lord, you will not be angry for asking questions."

"No, nobody is angry," Holt replied. "You may ask what questions you will."

Philip Middleton took the stand next. He spoke uninterrupted for around ten minutes, retracing the narrative that Dann had conveyed in his testimony. Middleton described the negotiations with Nicholas Trott in the Bahamas, and the bribe that the proprietor governor had accepted. At the end of his account, the prosecutors asked Middleton to affirm that all five of the accused men had arrived in Nassau with Every, and that all five had ultimately been welcomed by Trott. Taking over the prosecutorial reins from Newton, Holt had the witness review the distribution of stolen goods after the *Gunsway* heist. Perhaps explaining why he had turned against his shipmates, Middleton claimed that he had been granted a hundred pounds as his share of the loot, but that John Sparkes had subsequently stolen it from him.

The testimony of both Dann and Middleton made it clear that the state was using the second trial as a stage to denounce international piracy, despite the fact that the case itself involved the theft of a ship belonging to British citizens. Technically speaking, the attack on the *Gunsway* and the illicit negotiations with Trott had nothing to do with the mutiny that they were being accused of. And yet Newton and his allies on the bench spent hours of court time establishing the "rage and barbarity" of the men's crimes in the Indian Ocean, and the corruption of the colonial powers in the Bahamas. Had the defendants been granted actual legal representation, their lawyers would no doubt have objected to all the extraneous accounts of their Indian Ocean predations; these were crimes the men had already been acquitted of, after all. But common law courts were heavily weighted toward the authority of the state. The five men at the bar had no legal expertise in their corner, so Holt and Newton took the opportunity to drag all the offenses from the first trial onto the stage of the second.

"THE KING'S COUNSEL have done with the evidence," Holt announced to the defendants at the end of Middleton's testimony. "Now is your time for to speak, if you have any thing to say for yourselves."

In turn, each prisoner was allowed to make a statement or call back witnesses who had testified earlier. One by one, the prisoners made variations of the same argument: they had been forced into piracy against their will. Edward Forsyth asked for Thomas Druit to return to the witness box, and asked the former first mate of the *James* if he had not commanded him, on the night of the mutiny, to board the pinnace with the aim of fighting off the mutineers aboard the *Charles*.

"Yes, you were commanded," Druit acknowledged. "And then I commanded you to come back, and you refused."

"You did not command me back," Forsyth replied.

"Yes I did, and fired at you and shot through the boat."

Forsyth explained that his options were limited, once the other mutineers had joined him on the pinnace. "I held water with my oar. That was all I could do."

"Instead of rescuing the ship," Holt interjected, "you run away with her. He commanded you back and you refused to come back."

"I could not bring her back myself, nor come back, unless I should leap over board."

Holt asked Forsyth if he had anything else to say in his defense. Forsyth's final lines would be echoed by his fellow defendants: he had been swept up in the chaos of the mutiny and had been unable to return to the *James*.

"My lord," Forsyth began, "when I was in the boat. I knew not who was in it, nor how many. When I came aboard the Charles, the sails were loose and I was in a very sorry condition. They cut the boat off and put her in a drift. I could not get into her; she was gone in a minute's time. I did not know which way or what men there were in her, nor heard nothing til two o'clock the next day. And I hope, my Lord, as we are but poor sea-faring men, and do not understand the law, you will take it into consideration."

Holt bristled at Forsyth's profession of naiveté. "But all you seaman understand that law that it is not lawful to commit piracy," he snapped, "and he that doth deserves to be hanged."

Young William Bishop, James Lewes, and John Sparkes all relayed a similar narrative: commanded by Thomas Druit to man the pinnace, they had quickly found themselves overpowered by mutineers intent on joining their ringleader, Henry Every, aboard the *Charles*.

In his testimony, Bishop attempted to address one of the most damning elements from the original testimony: the undeniable fact that Every had allowed some of the sailors—including Gibson and Creagh—to leave the *Charles* of their own volition, in a boat that had room for many more.

"When we came aboard the ship Charles," Bishop explained, "they commanded the innocent to do what they pleased, with pistols and cutlasses, and they commanded me to go into the hold, to do what they pleased . . . And I heard afterwards, that none went ashore but whom they pleased—that is, Every and his crew. And I not knowing of it, could not go. And if I had known it, I had not been admitted to go."

THE MIDDLE-AGED STEWARD WILLIAM May attempted the most impassioned defense of the five prisoners. He began by stating that he knew nothing of the mutiny plot. "I believe very few knew of it; I believe not above nine or ten."

Holt dismissed the excuse out of hand: "None of them say you were at the consult. But one says that you said 'God damn you; you deserve to be shot through the head' and held a pistol to him."

"I never was any higher than the under deck," May protested. "I was coming up the hatch-way, and Captain Every was standing and commanding the ship."

"Every was no officer," Holt interjected. "He had nothing to command. He was under Captain Gibson, and took the ship from Gibson."

"My Lord, I knew nothing of the ship's going away," May pleaded.

"You should have stuck to Captain Gibson, and endeavored to suppress the insolence of Every. Captain Gibson was the commander. You ought to have obeyed him and if any had resisted him, or gone to put a force upon him, you should have stood by him."

"I was surprised," May responded weakly.

May then turned to question of the empty seats on the pinnace, and his ambiguous parting words with Gravet. The transcript conveys the sense of a man working as furiously as he can within the legal structure of the courtroom proceeding, trying turn the state's evidence against the prosecution, despite his limited knowledge of the law.

"When I came out again, they began to hurry the men away," May said. "Here was Mr. Gravet, the second Mate . . . I told him he should remember me to my wife, [as] I am not like to see her, for none could go, but who they pleased. For when those men were in the boat, they cried out to have a bucket or else they should sink, they having three leagues to go. And I do not know how they could go so far with more, when their boat was like to sink with those that were in her, as some of the King's evidence have testified."

Shortly thereafter, Judge Hedges intervened. "You seem to say that you were under a constraint and a terror," he said from the dais. "Did you make any complaint or discovery so soon as you had liberty, or at your first coming into the King's Dominion?"

May replied that he had "discovered" the whole affair to a magistrate in Rhode Island, before returning to England, and had been on his way to London to confess his crimes when he was apprehended. He then launched into a long story about his earlier illness during the voyage to the Red Sea. He reiterated his claim that he had missed significant stretches of the voyage, recovering from a fever on land. Every, May claimed, had attempted multiple times to get him back on the *Fancy*, but his illness had made it impossible.

"When Captain Every came in again," May protested, "I could not go nor stir."

"Do not call him Captain," Holt barked. "He was a pirate."

ONCE THE PRISONERS had concluded their pitiable attempts at self-defense, the solicitor general rose to make a closing argument for the prosecution. He began by reiterating the global implications of the verdict that Henry Newton had established at the outset of the first trial: "[The accused] could not find shelter in any other part of the world, and I hope you will make it appear such crimes shall not find shelter here . . . These are crimes against the laws of nations, and worse than robbery on land." He addressed the claims that the defendants had been overpowered by the mutineers, forced into service against their will: "Now they have only this to say for themselves: that they were forced to do what they did. But it has been proved to you that they were not forced. It was said: all might go that would."

Holt took over and reviewed the state's evidence one more time before commanding the jury to retire to its quarters and render a verdict. After a few hours, they returned with a question: Was there any evidence that John Sparkes in particular had consented to run away with the ship?

Holt dismissed the question briskly. "He was with them at the carrying off of the ship, and at the taking of the several prizes, and had his share afterwards. What is consent? Can men otherwise demonstrate their consent than by their actions?"

This simple question of consent—did William May and the others *choose* to join the mutiny, or were they compelled against their will—had truly momentous implications. It wasn't just a matter of life or death for the pirates. The entire objective of the trial—to establish to the world, and particularly to Aurangzeb, that England was at last determined to renounce piracy—would be sabotaged if the jury decided that May and the others were telling the truth about their opposition to the mutiny. Whatever schemes Annesley and Gayer had concocted to placate the Universe Conqueror back in India could be wiped away in an instant if the Lords Justices in Old Bailey failed to get a single conviction in back-to-back trials. Hedges, Holt, and their collaborators on the East India Company's special committee had done everything possible to create the ideal show trial to display the state's antipiracy stance. But now the plot of the show itself threatened to take a disastrous twist. One jury had already shrugged off the accusations of piracy. Could the second jury do the same to the charges of mutiny?

The jurors returned to their quarters and deliberated for a "very little time." When they emerged again, they stood together in the jury box. The clerk asked them if had reached a unanimous verdict. They answered in the affirmative.

"Ed Forsyth, hold up thy hand," the clerk instructed and then turned to the jury. "Look upon the prisoner. Is Edward Forsyth guilty of the piracy and robbery whereof he stands indicted, or not guilty?"

For each of the accused, the jury delivered its verdict. They were all guilty as charged.

Lord Justice Holt had the final word, as the convicted criminals were ushered out of the courtroom to await their sentencing in New-gate Prison. "Gentlemen," he said to the jury, "you have done extremely well, and you have done very much to regain the honor of the nation, and the city."

31

EXECUTION DOCK

The East End, London
November 25, 1696

Several days after the second trial ended, the six convicted mutineers—including Joseph Dawson, who had pled guilty twice—were brought back to Old Bailey for sentencing. Standing at the bar for one last time, each man was asked in turn by the clerk why they should not be sentenced to death for their crimes.

Dawson answered first, with an air of resignation: "I submit myself to the King and the honorable Bench." Forsyth maintained his innocence; the trial transcripts only note that he "went on to justify himself, etc." Judge Hedges intervened to explain that "the prisoners at the bar have had a very fair trial, and been fully heard upon your defense." But the jury's verdict has been rendered, the judge continued. The question now was whether the state had any reason not to execute the men for their capital crimes.

Giving up on his last-minute defense, Forsyth announced, "I desire to be sent to India to suffer there."

William May returned to his health complaints, and he, too, proposed an overseas assignment in lieu of execution. "My Lord, I being a very sickly man never acted in all the voyage," he protested. "I have served my King and Country this thirty years, and am very willing to serve in the East India Company where they please to command me, and desire the honorable Bench to consider my case, and if I must suffer, I desire to be sent into India to suffer there."

"I am an ignorant person," James Lewes conceded, "and leave myself to the King's mercy." John Sparkes requested the same mercy from the crown. Young William Bishop offered perhaps the most tragic of the statements. "I was forced away," he said, "and when I was but eighteen years old, and am now but twenty-one, and desire the mercy of the King and the Court."

The pleas for mercy went unanswered. Judge Hedges announced the sentences by steering the indictments back to the original international crimes of piracy, as though the acquittals of the original trial had never happened.

"You have been found guilty upon three several indictments, for the same detestable crimes committed upon the ships and goods of Indians, of Danes, and your own fellow-subjects," he announced. "The law for the heinousness of your crime hath appointed a severe punishment, by an ignominious death, and the judgement that the law awards is this: that you and everyone of you be taken from to the place from whence you came, and from thence to the place of execution, and there you, and every one of you be hanged by the necks, until you, and everyone of you be dead. And the Lord have mercy upon you." Only Joseph Dawson, who had pled guilty in both trials, was spared.

On November 25, 1696, two and a half years after they had—willingly or not—thrown in their lot with Bold Captain Every, the five men were lead out of Newgate Prison and marched through the streets down to the wharfs at Wapping on London's East End, not far from the dockyards where the *Charles II* had been originally built.

The specific location of Execution Dock is a matter of dispute among London historians. Three different pubs today claim to reside at the site of the original dock. But however ambiguous its exact location may be, we can imagine the general scene with some accuracy, given how closely public executions were covered by the tabloids of the day. The grim spectacle of the public hanging was, in a real sense, the closest equivalent in Every's age to the modern experience of major sporting events: an act of physical violence viewed live by teeming crowds and indirectly experienced by thousands thanks to media coverage.

Execution Dock faced the river, for symbolic reasons. The pirates who were hanged there—their bodies often left to decay for days—sent a message to the nautical community: *Do not delude yourself into thinking that you are beyond the reach of the law when you sail past the mouth of the Thames and into open water.* The riverside placement forced the audience to crowd into a flotilla of rowboats, anchored against the tide in front of the dock. Imagine all those prurient spectators, bobbing in the Thames for hours, waiting for the condemned to appear before them, eagerly anticipating the ritual sacrifice. Imagine a flurry of activity on the steps leading down from Wapping Street to the shoreline: the crowd lunging to its feet, full-throated, as the five prisoners made their way toward the gallows.

As in most public executions, there were last words to be delivered. On a bustling urban river, with no technical ability to amplify speech beyond the limits of human vocal cords, most of the words spoken by the prisoners went unheard by the throngs floating in the Thames. But they were quickly amplified textually by the press. Within a month, a pamphlet had been published, promising "An Account of the behaviour, dying speeches, and execution of . . . William May, John Sparcks, William Bishop, James Lewis, and Adam Foresith for robbery, piracy and felony, at the execution-dock."

Most of the confessions followed the conventional script of a crime-doesn't-pay morality play. Forsyth, for instance, observed that "besides the Guilt of his Offences, and the present capital Punishment, his Wicked Life, attended with many Hardships and Hazards he had undergone in his Robberies, was little less than a Punishment; for wickedness . . . brings great many troubles and afflictions along with it."

The last words of young John Sparkes, however, were the most haunting. He appeared to have been genuinely traumatized by sexual violence onboard the Mughal ship. "He expressed a due sense of his wicked Life," the pamphleteer reported, "in particular to the most horrid Barbarities that he had committed, which though upon the Persons of Heathen and Infidels, such as the forementioned poor Indians, so inhumanly rifled and treated so unmercifully; declaring that his Eyes were now open to his Crimes, and that he justly suffer'd Death for such Inhumanity, much more than his Injustice and Robbery, in Stealing and Running away with one of his Majesties Ships, which was of the two his lesser concern."

John Sparkes may have technically been convicted of mutiny aboard the *Charles II*, but he went to his grave atoning for the crimes he committed on the *Ganj-i-sawai*.

Their last words recorded, the five men stood on the gallows as a noose was tied around each of their necks. Pirates executed at the dock were subjected to an unusually cruel form of hanging, using a shorter rope than usual. The reduction in length meant that the neck would not break when the platform beneath them was pulled away. Enemies of all mankind did not deserve the split-second execution of a severed spinal cord. Instead, they were asphyxiated. Deprived of a sudden death, the five pirates dangled from the noose, their bodies twitching as they slowly suffocated in front of the jeering crowd.

With a guilty verdict and a public execution, the British government—and the East India Company—had managed to produce the show trial that they had originally planned. They had at last established a compelling ending for the dominant narrative. John Everingham's contract was renewed, and the printer released the court transcripts—with only a brief allusion to the unsuccessful first trial—as a twenty-eight-page bound volume within a matter of weeks. The publication went through multiple printings, and was read throughout the British empire. Its final lines left no doubt where the Crown stood regarding the crimes of piracy:

According to this sentence, Edward Foreseth and the rest were executed, on Wednesday, November the 25th, 1696; at Execution-Dock, that being the usual Place for the Execution of Pirates. FINIS.

EPILOGUE:
LIBERTALIA

A few days before his stolen coins were found quilted into his jacket—starting the whole chain of events that would lead to five of his shipmates hanging at Execution Dock—John Dann stumbled across Henry Adams's new bride in the London suburb of St. Albans. She was boarding a stagecoach by herself for an unknown destination. Dann and Mrs. Adams chatted briefly, and the former quartermaster's wife let it slip that she was headed off to meet with Henry Every.

The story of Dann's chance encounter with Mrs. Adams comes from Dann's original deposition, recorded by the authorities shortly after his arrest. If his account is true—and there is no apparent reason why Dann would fabricate such a detail—it paints an intriguing picture of Henry Adams's new bride. According to multiple testimonies, Adams had arrived in the Bahamas with the rest of the crew of the *Fancy*, and within a matter of weeks he had met and married the woman we know only as Mrs. Adams. Somehow, he had persuaded her to join him and nineteen of his mates for a two-thousand-mile journey to Ireland in a small sailboat. Against significant odds, they had made it to safe harbor, bribed their way past the landwaiter in

Dunfanaghy, and found themselves on British soil again, with their share of the treasure largely intact.

And then, after all that, Henry Adams and his wife had parted, leaving her boarding a stagecoach alone to meet with Captain Every at some undisclosed location. The fact that Dann seems to have had a civil conversation with her at St. Albans suggests that she had embarked willingly on the journey to Ireland, that her marriage to Henry Adams had been a voluntary one. But it is difficult to imagine the chain of circumstances that had left her, just a few weeks after arriving in Ireland, heading off for a secret rendezvous with her new husband's captain.

No doubt many innocuous explanations exist. Perhaps Adams and Every were together, and she just happened to forget to mention her husband's presence at the meeting. Perhaps Dann didn't bother to mention Adams to the magistrates interrogating him, assuming they were mostly interested in Every. Perhaps Dann made the whole thing up to get credit for pointing the state in the direction of their most wanted man. But if that was his strategy, then why not actually invent even more detail? Why wouldn't he have named a specific location for the rendezvous?

You can spiral down the chain of what-ifs for hours. But if you go with the simplest interpretation of Dann's testimony, there is something culturally suspicious—given the conventions of 1696 Britain—about Henry Adam's wife heading off to see Every by herself a month after marrying her husband. And that raises the question of whether a romantic entanglement was bringing Mrs. Adams to Captain Every. Did Every somehow steal his quartermaster's bride?

Whatever dim light it may shine on the romantic liaisons inside the Every gang, that brief section of Dann's testimony where he describes his meeting with Mrs. Adams has a significance that extends beyond the love lives of the pirates. Mrs. Adams's passing remark while boarding the stagecoach is the last legitimate trace of Henry Every's existence in the historical record. In early August 1696, a month after his return to Ireland, as Dann and Middleton were being interrogated, the most wanted man in the world simply disappeared. To this day, no one knows what happened to him.

HENRY EVERY THE MAN might have vanished in August 1696, but the mythological Henry Every would become increasingly visible over the next few decades. In 1709, Van Broeck published his mini biography, *The Life and Adventures of Captain John Avery*, written from the (almost certainly fictitious) vantage point of one of Every's crew. Van Broeck's account was the first published narrative to cast Every as a romantic suitor, dazzled by the tragic beauty of Aurangzeb's granddaughter. The story ends with Every and his bride happily settled in Madagascar, where Every established a thriving pirate kingdom. With a fleet of forty warships and fifteen thousand men, Captain Every—according to Van Broeck's account—appears to have had a second act as an urban planner: "Towns were built, communities established, fortifications built, and entrenchments flung up, as rendered his Dominions impregnable and inaccessible by sea and land."

In Van Broeck's hands, the story of Every had been transformed from the conventional pirate fantasy of the self-made man into something even more outlandish: an ascension, as Van Broeck put it, "from Cabin Boy to King." Bound up in that legend are two related utopian ideals that would have been mesmerizing to commoners back in England. First, the dream of extreme class mobility: that you could be born into a working-class family in Devonshire and, through the sheer force of your own daring and charisma, work your way not only into a vast fortune but also onto a royal throne, with thousands of loyal subjects and the granddaughter of the world's wealthiest man as your bride (albeit with somewhat strained relations with your in-laws). The second utopian impulse lies in the idea of a pirate kingdom itself: the egalitarian ethos of the pirate ship's collective brought ashore and rendered on a much larger scale.

The fantasy of the pirate utopia resonated so widely that the story reappeared in multiple formats throughout the early 1700s. The Theatre Royal in London staged a play called *The Successful Pyrate*, a somewhat slapstick account of Every's life, focusing on his years in his rogue state on Madagascar. In his bestselling *A General History of the Robberies and Murders of the Most Notorious Pyrates*, Charles Johnson interrupts the usual tales of daring raids on the high seas with surprisingly detailed accounts of a veritable pirate constitutional convention held on Madagascar. "The next day the whole colony was assembled," Johnson wrote, "and the three commanders proposed a form of a government, as necessary to their conservation . . . They looked upon a democratical form, where the people were themselves the makers and judges of their own laws, the most agreeable . . . The treasure and cattle they were masters of should be equally divided." Henry Newton had used the Every gang trial to assert a master narrative in which pirates were the enemies of all mankind. But in Johnson's rendition, the pirates become almost the mirror image: "Marine Heroes, the Scourge of Tyrants and Avarice, and the brave Asserters of Liberty." As one of the Madagascar settlers describes it, "They were not pirates, but men who were resolved to assert that liberty which God and nature gave them, and own no subjection to any, rather than was for the common good of all . . . [They] were vigilant guardians of the people's rights and liberties, [who] saw that justice was equally distributed." According to Johnson's account, the pirates gave their "democratical" state a name that would echo in the radical imagination back in Europe for years to come: Libertalia.

Every's crimes on the Indian Ocean ultimately helped define and fortify institutions that would come to dominate the modern world. Thanks to Samuel Annesley's ingenuity, the *Gunsway* affair would give the East India Company new powers that would ultimately lead to their imperial rule over the subcontinent; the contretemps with Aurangzeb forced the British government to clarify its long-ambiguous legal attitude toward piracy in international waters. Institutions like central governments or multinational corporations often seem as grand and formidable as the buildings they occupy. But the institutions themselves—and the power they wield—are invariably shaped by smaller conflagrations at their boundaries, defining the limits of their authority. Pirates occupied that role in the 1600s. Yet Every's story also lit a different fuse: the deeply populist vision of a society where the stratifications of wealth and privilege could be replaced by a much more equitable form of social organization. In time, that vision would lose its association with the sea dogs and the mutineers, and with Henry Every himself; pirates would be domesticated into children's books and theme park rides. But the pirate collective's radical dream of economic and political liberation would find new, more reputable vessels in the centuries to come.

To a certain extent, Every and the generation of pirates that followed him were drawn to those still nascent political structures by the intense challenges presented by the ocean itself. In those first centuries of the Age of Exploration, the ocean was a place that demanded constant experimentation. Life at sea is human culture at its most extreme, in one sense. You are surrounded by things that pose existential threats to you, thanks to your biology: water, thirst, starvation. And yet our cultural ingenuity gives us the opportunity to survive in such a hostile environment, even make a living from it. But humans had to invent new tricks to pull off such an impressive feat. Some of those tricks were technological: better maps and compasses and clocks. But some of them were political: new ways of organizing a polity, or distributing wealth.

We should not romanticize the populist strain that runs through Every's life as a pirate. The pirates helped cultivate a mythos—the underclass that fights its way to a more just society—that would be embraced by political progressives and revolutionaries endlessly over the centuries that followed. But those pirates were also, unquestionably, a gang of xenophobic sexual predators. The pirates tortured other human beings for purely mercenary ends. They burned a mosque as pointless act of retribution. They captured slaves and treated them as a kind of liquid currency instead of as human beings. They spent multiple days on a ship raping religious pilgrims. Karl Marx once said of capitalism that you had to think of it simultaneously as the best thing and the worst thing that had ever happened to human society. To make sense of the pirates—and of Henry Every most of all—we have to adopt a similar split consciousness. They were heroes to the masses. They were the vanguard of a new, more equitable and democratic social order. And they were killers and rapists and thieves, enemies of all mankind.

THE *GUNSWAY* CRISIS was ahead of its time in another way: in the asymmetrical relationship between the key actors, and the global scale of the event's ultimate consequences. One of the most striking things about the story of Every and his crew is the ability of such a small group of humans—working entirely outside the official institutions of power—to trigger events that would be heard around the world. The mix of fear, admiration, and disproportionate influence that Every unleashed on the planet represented a turning point in the evolution of the world system. It's a script we know by heart in the age of al-Qaeda and ISIS: rogue agents working outside the confines of traditional nation-states, using an act of violence to spark a geopolitical crisis and a global manhunt. But the first drafts of that script were written by Every and his men more than three centuries ago.

That watershed says less about Every himself than it does about the "new world order" that was coming into place in the late 1600s. The events reverberated so extensively not because of the power or cunning of a single individual, but rather because of the complex web of relationships that brought those two ships together that day in September 1695: the wealth of the Grand Mughals, the rising imperial ambitions of the British nation, the emerging importance of the nation-state, the birth of the modern multinational corporation, the increasingly vital networks of global trade, and the challenge to national borders and sovereignty that pirates posed. In a less-interconnected system, two hundred men would have had no way of triggering a truly global crisis with tangible effects on at least three different continents. Henry Every just happened to be one of the first to light up that network, and in lighting it up, Every and his men made it clear just how interdependent the whole system was, how easily it could be disturbed by seemingly minor players. The cannon explosion and the mainmast strike were a preview of Franz Ferdinand in the streets of Sarajevo: a single act of violence that threatens to set the world on fire.

Two months after the execution, James Houblon received his letter from the appalled Philadelphian, griping about the open display of the Every crew in the colonies. The letter is of historical interest in that it shines a light on the lax legal atmosphere for pirates in the colonies at the end of the seventeenth century. But it also illuminates another region on the map of Henry Every's life. According to the correspondent, Every's men claimed they "took an Indian princess captive, whom Every spirited away for himself, and left his men with several bags of gold apiece."

Perhaps these men were posers pretending to be part of Every's gang to impress the locals, drawing upon thirdhand mythology to make their stories more enticing. But the overheard conversation had to have happened by late 1696, probably within weeks of the November executions at Execution Dock—years before the heroic (and largely fictional) biographies would be published. Had the myth of Every and his Muslim bride already made it to Philadelphia through word-of-mouth channels? Or were those men in the Philadelphia pub telling the honest truth: they had been there in September 1695, off the coast of Surat, and they had seen with their own eyes Bold Captain Every "spirit away" his Indian princess. And if those are the facts, they raise an even more provocative question: Where did the Indian princess go?

Is it conceivable that the mysterious Mrs. Adams could have been the Indian princess, disguised for some reason for their entry into Ireland? Perhaps Every feared that her association with him—not to mention her ties to Aurangzeb—would make her an additional prize, were the authorities to capture Every himself. And so she took on a new identity, pretending to be his quartermaster's bride, and even parted ways with Every to keep up the illusion once they'd made it past the landwaiter, then waited until Every sent the signal to return to him.

But for that scenario even to be admissible into evidence, you have to accept at least some of the Van Broeck mythology of Every as the great romancer, which has significant plausibility problems. The princess, in that scenario, would have to be on some level willingly along for the ride, not a captive. Was it possible—given the reality of life circa 1695—for a well-to-do Muslim woman to see a band of pirates as an opportunity, as an escape route from the harem life awaiting her back in Delhi? Is there a more plausible version of Van Broeck's farcical account, with its instantaneous marriage proposal, where Every and the princess do form some sort of alliance? Perhaps Every's men do recognize her stature, and bring her to their captain, probably assuming that he will want her as a sexual conquest. It is apparent in the first seconds of the encounter that she belongs to a higher station than Every. She is the sophisticate, not the "noble savage." Every acknowledges the gap that separates them somehow. Perhaps he even tries to make the encounter into a lesson for his crew, given his well-documented concerns about their dangerous "hunger." Perhaps he does act with chivalry, as a message to his men. And the princess has her own complaints: she's a woman with a grandfather who happens to be the most orthodox Muslim in the four-hundred-year reign of the Mughal Dynasty. Perhaps she sees Every and the *Fancy* as her one chance to break free from her life of pampered oppression, not so much love at first sight as the lesser of two evils. And out of that small initial link a stronger bond is formed. She passes as Mrs. Adams and heads to London to lay low for a few weeks, and then, one day, the letter comes in the post and she catches that stagecoach to be reunited with Every.

The reality is probably much drearier or darker than the fabulist accounts of the pirate king or the great romancer. If the crew of the *Fancy* did, in fact, present an Indian princess to Every, he may well have "ravished" her in the criminal sense of the word. She would almost certainly not have spoken English, so whatever communication existed between them would have been extremely limited. And even if she had stayed on board the *Fancy* of her own volition, she might have died on the long voyage to the Bahamas. All we know for certain is the legend that developed around her: the pirate king and his Muslim bride. That legend itself is of historical interest, in the simple but remarkable fact that this rousing tale of a working-class hero—one of the first such tales to be amplified by the popular press—would end with a biracial marriage performed by a Muslim cleric. Today, of course, the idea of a working-class British man marrying a woman born into a well-to-do family from South Asia would be—in most circles—a routine matter, thanks to the global networks that were first coming into being in Every's time, and to the long battle against racism and religious intolerance that would be fought over the centuries that followed. But in Every's era, a multicultural romance of that sort was practically unheard of. And so, in the end, given the obscurity of her true biography, the meaning of the Indian princess has less to do with actuality and more to do with aspiration. It may not have been possible, in the last years of the seventeenth century, for a Muslim princess to run away with an English commoner and live happily ever after. But it was possible for an audience of readers to *want* that outcome, to imagine a world where such an alliance could be celebrated, not shunned.

The details of Henry Every's life—or at least the two-year stretch of it where he became a pirate—were far better documented than the life of the Indian princess. But the end of his life remains equally mysterious. No plausible evidence suggests that Every made it back to Madagascar, and Libertalia itself appears to have largely been a fantasy, spun by the balladmongers and scribes dreaming of a better way back in class-stratified London. (When Woodes Rogers visited the pirate community on Madagascar in 1710, he found that the population had "dwindled to between 60 and 70, most of them very poor and despicable, even to the Natives, among whom they had married.") According to Charles Johnson, Every lost most of his fortune trying to launder it through gold dealers shortly after his return. In that version of the story, he died in poverty—and, almost as amazingly, in obscurity—in Devon, twenty years after his return.

But the truth is, no one really knows what happened to Henry Every. He had the full glare of a global spotlight on him for a few years, and somehow he snuck back into the shadows.

SAMUEL ANNESLEY'S PLAN for a "salt-water *faujdar*" suffered a few major setbacks in its first years of implementation. One of the first captains hired by the East India Company to protect the Mughal fleet turned out to be William Kidd, who set off for the Indian Ocean in 1696 in a brand-new, thirty-four-gun ship, the *Adventure Galley*. Kidd's official mission turned out to be more challenging that he had originally anticipated, and he soon converted to piracy himself, at one point capturing an Indian Armenian ship called the *Quedagh Merchant*, provoking almost as much outrage as the *Gunsway* affair. Kidd proved to be less skillful at evading capture than Henry Every, however; he was arrested in Boston several years later and ultimately sent back to England for trial. He was hung at Execution Dock in May 1701, five years after the Every gang met the same fate.

Kidd is generally considered the last of the "Red Sea Men"; with Mughal and merchant ships under increasingly effective protection from the English *faudjar*, the pirates shifted focus to the Caribbean and left India to the legitimate traders. (Young Philip Middleton, who had testified against his former mates in the Old Bailey trial, became one of those traders, working for the East India Company in Bengal.) In the decades that followed, the military force that Annesley had first imagined while chained in irons in the Surat factory grew increasingly central to the company's presence in India. By the 1750s, the company maintained a small army of three thousand troops on the subcontinent; by the 1800s, they would number in the hundreds of thousands.

Aurangzeb would go on to outlive many of his descendants,

dying in 1707 at the age of eighty-nine. In his final years, the Universe Conqueror sensed that the Mughal dynasty was on unstable ground. "After me, chaos," he is said to have predicted. It turned out to be an accurate forecast. For fifty years after his death, the Indian state was characterized by a "a string of weak emperors, wars of succession, and coups by noblemen." All the while, the East India Company consolidated its power over the region, culminating in the Battle of Plassey in 1757, after which the corporation assumed official control of the subcontinent, an administrative reign that would last for a hundred years.

If Samuel Annesley had played an important role in that rise to power, his contributions were hardly recognized by the company he had spent the first few decades of his adult life serving. Shortly after the Every affair, he was fired for allegedly mismanaging the accounts at the Surat factory. He remained in Surat as a private trader, with varying success, and lived until the then-impressive age of seventy-seven. Near the end of his life, he fell on hard times and began plotting a return to England. "Such a continual succession of troubles in an unhealthy climate," he wrote in a letter, "makes me rather desirous of a quiet retreat in my native country than to continue any longer in India." But it was too late to arrange the voyage back. He died in Surat in 1732. After his initial voyage to India at the age of nineteen, Annesley never saw England again.

DID HENRY NEWTON'S MASTER narrative—England's formal renunciation of piracy—ultimately win out, despite the disaster of the first Every trial? In the long run, yes. Every certainly inspired the Golden Age pirates that terrorized the Caribbean in the early 1700s, but from the Old Bailey trial on, the British government adopted a unified front about the legal status of the pirates. Newton's opening line from the first trial—"Suffer pirates, and the commerce of the world must cease"—became a core principle. John Gayer's pleas from Bombay Castle were turned into official state policy. The East India Company's trade with the Mughal empire and the merchants of Surat recovered, particularly after the execution of William Kidd. With East Indiamen patrolling the waters, Red Sea piracy declined, and the pilgrim ships once again made their journey to the hajj unmolested. England—and her colonies—slowly shed their reputation as a nation of pirates.

In March 1701, King William III released a "Proclamation for the Apprehension Of Pirates," reiterating the firm antipiracy position they had attempted to dramatize with the Old Bailey trial and borrowing some of the techniques that they had used in the initial bounty that had been put on Every's head. Any pirates who turned against their former shipmates and reported them to the authorities would "receive our Most Gracious Pardon for the Piracies before that committed by him"—along with a third of any treasure seized by the state based on their information. Any British subject living outside the law as a pirate could receive full clemency—and a meaningful reward—by turning in one of his fellow criminals.

The proclamation ran on for several pages, with the usual ornate qualifications and stylistic flourishes typical of such documents. But the very last line contained a striking proviso. The offer of clemency applied to every single royal subject who had turned to piracy, except one: Henry Every.

Acknowledgments

Almost a decade and a half ago, I published a book called *The Ghost Map*, about a cholera epidemic in London in 1854. Like most of my books it jumped across multiple disciplines—from microbiology to urban planning to sociology. But, unlike most of my books, it had what the Hollywood people call a "through-line": a central, more or less linear, narrative that the book rarely strayed far from. There was a killer loose on the streets of London and a (medical) detective on the case. Wherever the commentary happened to take you, you never deviated far from that main arc.

Somehow, despite its grim subject matter—or perhaps because of it—*The Ghost Map* slowly developed a meaningful readership over the years since it was published. And so in a way, the idea for this book first came out of the many conversations I had with the readers of *The Ghost Map* during that period. There was something in the structure of that book that drew readers in, something that kept them turning the pages. This book was an attempt to get back to a variation on that structure after a long absence: in this case, a pirate loose on the open sea, and an entire planet trying to find him. So it seems right to begin an acknowledgments by thanking those readers for reminding me how fun it is to write a book with single thread.

The story of Henry Every is similar to that of *The Ghost Map* in that it has been the subject of much academic scrutiny while remaining a story that most lay readers know close to nothing about. That is a particularly fortuitous place to find yourself as an author, building on a foundation of exemplary scholarship. I have tried to pay tribute to that work and commentary in the preceding pages—to make this a book not just about what happened in the story of Henry Every and the *Fancy*, but also a book about the *debate* about what happened. So a special thanks is warranted to the scholars and friends who shaped this book with their work, suggestions, and in some cases close reading, particularly Philip J. Stern, Douglas R. Burgess, David Olusoga, Joel Baer, Soma Mukherjee, Chris Himes, Mark Bailey, Stewart Brand, and Adam Fisher. I would also like to acknowledge my long-ago mentor from graduate school, Edward Said, who first got me thinking about how much the institutions of the West were shaped by their encounters with the "Orient." I wish he were still around to read this book, if only to see that I finally managed to get rid of most of the poststructuralist jargon that used to annoy him so much back in those grad school days.

I am also grateful to Joe Davies for some last minute research deep in the stacks in London. Thanks as well to the many institutions whose archives were essential to writing this book: the India Office Records at the British Library; the National Archives of the United Kingdom; the National Maritime Museum; the Docklands Museum; and the New York Public Library.

My editor Courtney Young was so devoted to figuring out the best structure for this book that she briefly converted her workspace into one of those "crazy-walls" from films like *The Usual Suspects*, visualizing all the different ways the chapters could be organized. She helped me work through some of the thorniest issues that the subject matter presented and kept an invaluable eye on the narrative pacing of the book. Once again, Kevin Murphy did a masterful job shepherding this book from an unruly early manuscript into a much more presentable form. My longtime publisher Geoffrey Kloske was as creative and flexible as always in getting this book to print. With one exception, my partnership with Geoff has lasted longer than any other in my publishing career: here's hoping that we have many more productive years together.

That one exception is my agent, Lydia Wills, who has worked with me for almost twenty-five years now. This is the thirteenth book that we have worked on together. That number should speak for itself, but in case it doesn't: Lydia has been a guiding light for my entire career, the person who got me thinking about the long arc of a "career" in the first place. I'm also delighted to have had help with this project and others from the good folks at Endeavor, particularly Ari Emmanuel, Jay Mandel, Sylvie Rabineau, and Ryan Mc-Neily.

I am grateful to my family for tolerating my long—occasionally interesting—digressions about seventeenth-century piracy over dinner. (Special thanks to my son Dean for suggesting "Libertalia" as a chapter title.) This book is dedicated to my wife, Alexa Robinson, whose lifelong passion for nautical history made her an ideal early research assistant for this book, before more important matters took precedent. And as always, she was a brilliant line editor, even if she took a little too much pleasure in mocking my "landlubber" phrasings at a few points. This one is for you, Lexie.

Marin County, CA
July 2019

Notes

INTRODUCTION

3 how to make an explosion: Parker, p. 63.

5 Some of the first differential equations: Steele, p. 360.

1. ORIGIN STORIES

14 "They found seldom less aboard": Turley, p. 23.

15 "Wandering scholars seeking alms": Quoted in Dean, p. 60.

17 Given that this memoir: Defoe, *The King of Pirates*, loc. 65–67.

2. THE USES OF TERROR

22 "They were dragged": D'Amato, loc. 1095–1097.

22 "His majesty is gone forth": Egerton and Wilson, plates 37–39, lines 8–23.

23 While they lacked the armies: See Hitchcock and Maeir for a nuanced comparison of the Sea Peoples and the pirates of the "Golden Age."

24 "terrorism and not royalty": Quoted in https://founders.archives.gov /documents/Jefferson/01-28-02-0305.

24 In a letter written just: For more on the evolution of terrorism, see https://www.merriam-webster.com/words-at-play/history-of-the -word-terrorism.

27 François L'Ollonais was reported: Leeson, pp. 113–14.

27 In one later version: Ibid., p. 112.

27 *Unparallel'd Cruelty:* The full description bears repeating, if only to remind the modern reader that tales of seemingly gratuitous violence have a long history: "After whipping [the boy], [he] pickled him in Brine; that for nine Days and Nights he tied him to the main Mast, his Arms and Legs being

all the Time extended at full Length; that not content with this, he had him unty'd, and laid along upon the Gangway, where he trod upon him, and would have had the Men done the same, which they refus'd; by which being exasperated as thinking, which indeed he might very well do, that they pitied him, he kick'd him about as he lay, unable to get up, and stamp'd upon his breast so violently, that his Excrement came up involuntarily from him; which he took up, and with his own Hands forc'd it several Times down his Throat; that the poor miserable Creature was eighteen Days a dying, being cruelly allowed Food enough to sustain Life, and keep him in Torture all that Time; that he was severely whipp'd every Day, and particularly the Day he died; that when he was in the Agonies of Death, and speechless, his inexorable Master gave him eighteen Lashes; that when he was just expiring, he put his Finger to his Mouth, which was took for a Signal of desiring something to drink, when the Brute, to continue his Inhumanity to the last, went into the Cabbin for a Glass, which he pissed in, and then gave it him for a Cordial; that a little, 'twas believed, went down his Throat; upon which pushing the Glass from him, he that Instant breathed his last; and God in Mercy put an End to his Sufferings, which seemed to cause an Uneasiness to the Captain for not continuing longer." Quoted in Turley, pp. 10–11.

28 "To prevent captives from withholding": Leeson, pp. 111–12.

3. THE RISE OF THE MUGHALS

30 "Should there be rain": Anonymous, "The Bolan Pass," pp. 109–12.

32 From 1 CE to 1500 CE: Maddison, loc. 7583–7584.

32 "which produce a kind of wool": Quoted in Yafa.

33 What made Indian cotton unique: For more on the world-historical impact of dyed fabric—valuable purely for its aesthetic properties—see Johnson, *Wonderland*, pp. 17–30.

33 The result was a fabric: Yafa, p. 28.

33 "There were in India trees": Strabo's writings on India are excerpted at https://www.ibiblio.org/britishraj/Jackson9/chapter01.html.

33 the Greek historian Strabo: "Wine, bronze, tin, gold, and various manufactured articles were shipped up the Nile to Coptos and moved overland to Red Sea ports at Myos Hormos or Berenice. Manned by Egyptian Greeks, they sailed through the Gulf of Aden to India by two major routes—to the north around Gujarat and to the southwest coast at Kerala or further south to Ceylon (see Casson 1989). They brought back spices, pepper, jewellery, and cotton goods. They were able to buy

Chinese silks, mirrors, and other goods which had come overland to India. The Indian trade was financed in substantial part by export of silver and gold. The volume and dating of Roman coins found in India provide an indication of the locus and changing intensity of commerce." Maddison, loc. 3884–3891.

35 "They shall eat every fourth": Quoted in Gopalakrishnan, 2008.

35 "The Hindus believe": Al-Biruni, pp. 10–11.

36 Three years later: "In 1012 it carried him to Thanesar, Harsha's original capital due north of Delhi. Anandapala, whose kingdom was now reduced to a small corner of the eastern Panjab and whose status was little better than that of a Ghaznavid feudatory, tried to intercede. He offered to buy off Mahmud with elephants, jewels and a fixed annual tribute. The offer was refused, Thanesar duly fell, and 'the Sultan returned home with plunder that it is impossible to recount.' 'Praise be to God, the protector of the world for the honour he bestows upon Islam and Musulmans,' wrote al-Utbi." Keay, loc. 4472–4476.

36 Mahmud's greed: In 1018, Mahmud's forces reached the sacred temple of Mathur, which they promptly "burned with naphtha and fire and levelled with the ground." An even worse fate awaited the temple of Somnath, near the coast of the Saurashtra peninsula. "After stripping it of its gold," the historian John Kealy writes, "he personally laid into it with his 'sword'—which must have been more like a sledgehammer. The bits were then sent back to Ghazni and incorporated into the steps of its new Jami Masjid (Friday Mosque), there to be humiliatingly trampled and perpetually defiled by the feet of the Muslim faithful." Kealy, loc. 4456.

36 If ever there were an uprising: Braudel, p. 232.

4. HOSTIS HUMANI GENERIS

40 in Van Broeck's account: Van Broeck, pp. 3–4.

43 "In return for this legal protection": Konstam, loc. 553–558.

43 "Privateers in time of War": Johnson, *A General History of the Pyrates*, p. 2.

43 "legitimate trade, aggressive mercantilism": Burgess, pp. 21–22.

44 "Drake's colossal success": Ibid., pp. 27–28.

5. TWO KINDS OF TREASURE

46 "such liberties of traffique": Foster, p. 61.

46 "I told him that my comming": Ibid., p. 82.

48 "Of chaires of estate": Ibid., p. 102.

49 "He is exceeding rich": Ibid., p. 104.

50 "India had long been": Keay, loc. 6673–6684.

52 "earls and dukes, privy councilors": Baladouni, p. 66.

6. SPANISH EXPEDITION SHIPPING

56 "I have nowhere met": Quoted in Charles Rivers Editors, loc. 28.

58 In August 1693: There is some ambiguity in the historical record over when and where the Expedition left England. The historian of piracy Angus Konstam, for instance, has them departing from Bristol in June, not London in August. See Konstam, loc. 4290–4291.

59 At the other end: "Based on a sample of 169 early-eighteenth-century pirates Marcus Rediker compiled, the average pirate was 28.2 years old. The youngest pirate in this sample was only 14 and the oldest 50— ancient by eighteenth-century seafaring standards. Most pirates, however, were in their mid-twenties; 57 percent of those in Rediker's sample were between 20 and 30. These data suggest a youthful pirate society with a few older, hopefully wiser, members and a few barely more than children. In addition to being very young, pirate society was also very male. We know of only four women active among eighteenth-century pirates." Leeson, p. 10.

7. THE UNIVERSE CONQUEROR

63 "much derangement": Keay, p. 214.

64 "the essential nature of Hinduism": J. F. Richards, *The Mughal Empire*, p. 152.

64 According to Khafi Khan: Ibid., p. 223.

65 "Beholding the dispersion": Ibid., p. 224.

66 "Mountain after mountain": Ibid., p. 244.

8. HOLDING PATTERNS

70 "We constantly drank our urine": Turley, p. 16.

70 "For we had nothing": Ibid., pp. 17–18.

71 "Take a hard egg": Quoted in Preston, pp. 29–30.

72 By the time he returned: Turley, p. 14.

9. THE DRUNKEN BOATSWAIN

79 "Is the drunken boatswain": All direct quotations from the mutiny are taken from the transcript of the Every gang trial, published by Everingham in 1696.

10. THE *FANCY*

89 "I. Every man shall": Johnson, *A General History of the Pyrates*, p. 116.

91 "the supream Power": Leeson, p. 29.

92 "For the Punishment of small Offences": Johnson, *A General History of the Pyrates*, p. 213.

93 According to eighteenth-century pirate: Leeson, pp. 59–60.

94 J. S. Bromley wrote: Quoted in Baer, "Bold Captain Avery," p. 13.

94 "Pirates constructed a culture": Redicker, *Between the Devil and the Deep Blue Sea*, p. 286.

94 They were also advancing populist values: These egalitarian values extended to the everyday interactions onboard. According to Johnson's *General History*, "Every Man, as the Humour takes him . . . [may] intrude [into the captain's] Apartment, swear at him, seize a part of his Victuals and Drink, if they like it, without his offering to find Fault or contest it." Johnson, *A General History of the Pyrates*, p. 180.

11. THE PIRATE VERSES

98 "Being An Account of *John Jewster*": These quotations are from the English Broadside Ballad Archive, maintained at http://ebba.english .ucsb.edu. Note that "Murderers Lamentation" uses the archaic "Mutherers" word in the original. I have translated it into the modern word "murderer" for legibility here.

100 "Our Names shall be": This is the ballad in its entirety:

> Come all you brave Boys, whose Courage is bold,
> Will you venture with me, I'll glut you with Gold?
> Make haste unto Corona, a Ship you will find,
> That's called the Fancy, will pleasure your mind.
>
> Captain Every is in her, and calls her his own;
> He will box her about, Boys, before he has done:
> French, Spaniard and Portuguese, the Heathen likewise,
> He has made a War with them until that he dies.
>
> Her Model's like Wax, and she sails like the Wind,
> She is rigged and fitted and curiously trimm'd,
> And all things convenient has for his design;
> God bless his poor Fancy, she's bound for the Mine.

Farewel, fair Plimouth, and Cat-down be damn'd,
I once was Part-owner of most of that Land;
But as I am disown'd, so I'll abdicate
My Person from England to attend on my Fate.

Then away from this Climate and temperate Zone,
To one that's more torrid, you'll hear I am gone,
With an hundred and fifty brave Sparks of this Age,
Who are fully resolved their Foes to engage.

These Northern Parts are not thrifty for me,
I'll rise the Anterhise, that some Men shall see
I am not afraid to let the World know,
That to the South-Seas and to Persia I'll go.

Our Names shall be blazed and spread in the Sky,
And many brave Places I hope to descry,
Where never a French man e'er yet has been,
Nor any proud Dutch man can say he has seen.

My Commission is large, and I made it my self,
And the Capston shall stretch it full larger by half;
It was dated in Corona, believe it, my Friend,
From the Year Ninety three, unto the World's end.

I Honour St. George, and his Colours I were,
Good Quarters I give, but no Nation I spare,
The World must assist me with what I do want,
I'll give them my Bill, when my Money is scant.

Now this I do say and solemnly swear,
He that strikes to St. George the better shall fare;
But he that refuses, shall sudenly spy
Strange Colours abroad of my Fancy to fly.

Four Chiviligies of Gold in a bloody Field,
Environ'd with green, now this is my Shield;
Yet call out for Quarter, before you do see
A bloody Flag out, which our Decree,

No Quarters to give, no Quarters to take,
We save nothing living, alas 'tis too late;
For we are now sworn by the Bread and the Wine,
More serious we are than any Divine.

Now this is the Course I intend for to steer;
My false-hearted Nation, to you I declare,
I have done thee no wrong, thou must me forgive,
The Sword shall maintain me as long as I live.

12. DOES SIR JOSIAH SELL OR BUY?

102 The Spanish Expedition investors: Baer, "Bold Captain Every," p. 12.

105 The wealth creation: Robins, pp. 48–49.

105 "The East India Stock": Defoe, *Anatomy of Exchange Alley*, p. 14.

108 "The mint was there": Wright, p. 38.

108 "marble seraglios": Ibid., p. 101.

108 "Was it not on account": Ibid., p. 112.

109 "Though our business": Quoted in Keay, *The Honourable Company*, loc. 2627–2629.

109 "pestilential vapours": Wright, p. 103.

109 "behave themselves": Keay, *The Honourable Company*, loc. 2713.

109 "Even by the lax standards": Robins, p. 54.

13. WEST WIND DRIFT

118 "forty pounds of gold": Baer, *Pirates of the British Isles*, pp. 86–97.

15. THE AMITY RETURNS

128 During the golden age: Leeson, p. 9.

16. SHE FEARS NOT WHO FOLLOWS HER

131 "Your Honor's ships": Quoted in Earle, p. 129.

132 "shrewd tactic to avoid": Baer, *Pirates of the British Isles*, p. 98.

17. THE PRINCESS

137 "Little did the unhappy": Bernier, pp. 13–14.

137 "nine youths": Mukherjee, p. 19.

138 "These ladies lived": Ibid., p. 1.

18. THE *FATH MAHMAMADI*

145 "drove a trade equal": Charles River Editors, loc. 280–284.

149 And they had exceeding: Van Broeck has a more romantic interpretation of Every's daring: "He exerted such a Courage, as if he had

prophetically known that the Reward of his Victory should be the most charming of the fair Sex, and the most previous of all inestimable Things, that the East can present us." Van Broeck, p. 29.

19. EXCEEDING TREASURE

152 "It is known that the Eastern People": Johnson, *A General History of the Pyrates*, p. 12.

153 Married in front of a Muslim: Van Broeck's account went on to stress that the remainder of the crew behaved with comparable chivalry: "The rest of the crew then drew lots for her servants and, to follow the example of their commander, even stay'd their stomachs 'till the same priest had said Grace for them." Van Broeck, p. 29.

153 "Such a Sight of Glory": Defoe, *The King of Pirates*, loc. 723–726.

153 "There was one of her Ladies": Ibid., loc. 733–737.

153 "carried off as captive": Wright, p. 160.

154 "If any of the Princess's Women": Defoe, *The King of Pirates*, loc. 912–914.

20. THE COUNTER NARRATIVE

161 "She they abused very much": Charles River Editors, loc. 311–316.

162 "the inhuman treatment": Quoted in Grey, p. 45.

21. VENGEANCE

167 "The town is so defiled": Stern, *The Company-State*, p. 134.

167 "the libertye of a Penn": Ibid., p. 135.

168 "It is needless to write": Keay, *The Honourable Company*, loc. 3485.

168 "For nine years past": Quoted in Wright, p. 168.

169 "The total revenue": Elliot, p. 354.

169 "To such a fanatical": Wright, p. 174.

22. A COMPANY AT WAR

173 "When the East India ships": Quoted in Robbins, p. 55.

179 the most important line: Proclamation for apprehending pirates, July 17, 1696, H/36 ff. 201–3.

23. THE GETAWAY

186 Reverting to his Benjamin Bridgeman alias: Baer, p. 103.

187 "They were afraid": Narrative of Philip Middleton, TNA/CO 323/3.

187 "But Captain [Every] withstood": Examination of John Dann, TNA/CO 323/2/24.

25. SUPPOSITION IS NOT PROOF

197 "How could I know?": Fortescue, p. 507.

26. THE SALTWATER *FAUJDAR*

198 "find out the pyrates": Govil, p. 410.
199 "As his land *faujdars*": Wright, p. 178.
201 "renewal of warfare": Ibid., p. 176.

27. HOMECOMINGS

205 a letter from a scandalized colonist: Burgess, p. 911.
206 "I heard he went over": The original deposition of Dann is recorded in the third person; for clarity, I have translated his account into a first-person voice. Examination of John Dann, TNA/CO 323/2/24.
206 "Captain Henry Every now goes by the name": Blackbourne to Chester, IOR H/36 f. 195–96.
207 The special committee spent: Stern, *The Company-State*, pp. 138–39.
207 They even covered the cost: Court Minutes, 19 IOR B/41 ff. 86, 97, 252.

28. A NATION OF PIRATES

212 "It placed the Every pirates": Burgess, p. 894.
215 Subsequent court committee minutes: Court Minutes, IOR B/41 f. 105, 143.

29. THE GHOST TRIAL

221 He then relayed: Newton's opening statement is worth reading in depth: "And the end was suitable to their beginning, they first practiced their crimes upon their own Country-men, the English, then continued them onto strangers and foreigners: for the ship in which this piracy was committed was an English vessel, called the Charles the Second, belonged to several merchants of this city, designed for other ends and a far different voyage, which these criminals, with the assistance of one Every, their Captain, in all these villanies, was seized near the Groyn in Spain, in May 1694, from which place, having first by force set Captain Gibson the Commander on shore, they carried off the ship, and with it committed many and great piracies, for several years (as will appear in the course of evidence) in most parts of the known world, without distinction upon all nations, and persons of all religions."

221 "Piracy . . . by so much exceeds": All quotations from both trials are from the Everingham transcript, published shortly after the execution of the Every gang in late 1696. In some cases, I have changed the spelling or grammar to modern usage for clarity.

30. WHAT IS CONSENT?

226 "brilliant act of legal legerdemain": Burgess, p. 901.

EPILOGUE

246 "The next day the whole colony": Johnson, *A General History of the Pyrates*, p. 432.

246 "They were not pirates": Ibid., p. 389.

252 In that version: "Whether he was frightened by these Menaces, or had seen some Body else he thought knew him, is not known; but he went immediately over to Ireland, and from thence sollicited his Merchants very hard for a Supply, but to no Purpose, for he was even reduced to beggary: In this Extremity he was resolved to return and cast himself upon them, let the Consequence be what it would. He put himself on Board a trading Vessel, and work'd his Passage over to Plymouth, from whence he travelled on Foot to Biddiford, where he had been but a few Days before he fell sick and died; not being worth as much as would buy him a Coffin." Johnson, *A General History of the Pyrates*, p. 15.

Bibliography

Al-Biruni. *India*. New Delhi: National Book Trust, 2015.

Anonymous, "The Bolan Pass." *Journal of the Royal Geographical Society of London* 12 (1842), 109–12.

Baer, Joel. *Pirates of the British Isles*. Stroud UK: Tempus, 2005.

————. "Bold Captain Avery in the Privy Council: Variants of a Broadside Ballad from the Pepys Collection." *Folk Music Journal* 7, no. 1 (1995), 4–26.

————. "William Dampier at the Crossroads." *International Journal of Maritime History* VIII, no. 2 (December 1996), 97–117.

Baladouni, Vahé. "Accounting in the Early Years of the East India Company." *The Accounting Historians Journal* 10, no. 2 (Fall 1983), 64–68.

Bernier, Francois. *Travels in the Mogul Empire (1656–1668)*. New Dehli: Oriental Books Reprint Corporation, 1983.

Best, Thomas. *The Voyage of Thomas Best to the East Indies (1612–1614)*. William Foster, ed. London: The Hakluyt Society, 1934.

Bialuschewski, Arne. "Black People under the Black Flag: Piracy and the Slave Trade on the West Coast of Africa, 1718–1723." *Slavery and Abolition* 29, no. 4, 461–75.

Braudel, Fernand. *A History of Civilizations*. New York: Penguin, 1988.

Burgess Jr., Douglas R. "Piracy in the Public Sphere: The Henry Every Trials and the Battle for Meaning in Seventeenth-Century Print Culture." *Journal of British Studies* 48, no. 4 (Oct. 2009), 887–913.

————. *The Pirates' Pact: The Secret Alliances Between History's Most Notorious Buccaneers and Colonial America*. New York: McGraw-Hill Education, 2008. Kindle Edition.

Casey, Lee A. "Pirate Constitutionalism: An Essay in Self-Government." *Journal of Law and Politics* 8 (1992), 477.

Charles River Editors. *Legendary Pirates: The Life and Legacy of Henry Every.* Charles River Editors, 2013. Kindle Edition.

Cordingly, David. *Under the Black Flag: The Romance and the Reality of Life Among the Pirates.* New York: Random House, 2013.

D'Amato, Raffaele. *Sea Peoples of the Bronze Age Mediterranean c.1400 BC—1000 BC.* London: Bloomsbury Publishing, 2015. Kindle Edition.

Dean, Mitchell. *The Constitution of Poverty: Towards a Genealogy of Liberal Governance.* London: Routledge, 2013.

Defoe, Daniel. *Anatomy of Exchange Alley: or A System of Stock-Jobbing.* London: E. Smith, 1719.

————. *The King of Pirates: Being an Account of the Famous Enterprises of Captain Avery, the Mock King of Madagascar.* Kindle Edition.

Earle, Peter. *The Pirate Wars.* New York: Macmillan, 2013.

Edgerton, William F., and John A. Wilson. *Historical Records of Ramses III: The Texts in Medinet Habu, volumes I and II.* Chicago: University of Chicago Press, 1936.

Elliot, H. M. *The History of India, as Told by Its Own Historians. The Muhammadan Period, Vol. 7.* London: Trübner & Co., 1871.

Emsley, Clive, et al. "Historical Background—History of The Old Bailey Courthouse." Old Bailey Proceedings Online, www.oldbaileyon line.org.

Examination of John Dann, 10 August 1696, The National Archives (TNA): Public Record Office (PRO) Colonial Office (CO) 323/2/24.

Findly, Ellison B. "The Capture of Maryam-uz-Zamānī's Ship: Mughal Women and European Traders." *Journal of the American Oriental Society* 108, no. 2 (Apr.–Jun. 1988), 227–238.

Fortescue, J. W., ed. *Calendar of State Papers, Colonial Series.* London: Mackie and Co., 1905.

Foster, William, Sir. *Early Travels in India, 1583–1619.* London: Oxford University Press, 1921.

Gopalakrishnan, Vrindavanam S. "Crossing the Ocean." *Hinduism Today.* July 2008. https://www.hinduismtoday.com/modules/smartsection/item .php? itemid=3065.

Gosse, Philip. *The History of Piracy.* Mineola, NY: Dover Maritime, 2012. Kindle Edition.

Govil, Aditi. "Mughal Perception of English Piracy: Khafi Khan's Account of the Plunder of 'Ganj-i-Sawai' and the Negotiations at Bombay." *Proceedings of the Indian History Congress* 61, Part One: Millennium (2000–2001), 407–12.

Grey, Charles. *Pirates of the Eastern Seas*. London: S. Low, Marston, and Co., 1933.

Hanna, Mark G. *Pirate Nests and the Rise of the British Empire, 1570–1740*. Chapel Hill: Omohundro Institute and University of North Carolina Press, 2017. Kindle Edition.

Hitchcock, Louise and Maeir, Aren. "Yo-ho, yo-ho, a seren's life for me!" *World Archaeology* 46, no. 4 (June 2014), 624–64.

Houblon, Lady Alice Archer. *The Houblon Family: Its Story and Times*. London: Archibald Constable and Company, 1907.

John, Ian S. *The Making of the Raj: India Under the East India Company*. Santa Barbara: ABC-CLIO, 2012.

Johnson, Charles. *A General History of the Pyrates*. Manuel Schonhorn, ed. Mineola, NY: Dover, 1999.

Johnson, Steven. *Wonderland: How Play Made the Modern World*. New York: Riverhead, 2016.

Keay, John. *India: A History*. New York: HarperCollins Publishers, 2010. Kindle Edition.

———. *The Honorable Company: A History of the East India Company*. New York: Harper Collins, 2014. Kindle Edition.

Khan, Iftikhar Ahmad. "The Indian Ship-Owners of Surat in the Seventeenth Century." *Journal of the Pakistan Historical Society* 61, no. 2 (April 2013).

Konstam, Angus. *Pirates: The Complete History from 1300 BC to the Present Day*. Guilford, CT: Lyons Press, 2008. Kindle Edition.

Lane, Kris. *Pillaging the Empire: Global Piracy on the High Seas, 1500–1750*. London: Routledge, 1998. Kindle Edition.

Leeson, Peter T. *The Invisible Hook: The Hidden Economics of Pirates*. Princeton, NJ: Princeton University Press, 2009. Kindle Edition.

Maddison, Angus. *Class Structure and Economic Growth: India and Pakistan Since the Moghuls*. London: Routledge, 2013.

———. *Contours of the World Economy 1–2030 AD: Essays in Macro-Economic History*. Oxford, UK: Oxford University Press, 2007. Kindle Edition.

Mukherjee, Soma. *Royal Mughal Ladies: And Their Contribution*. New Dehli: Gyan Publishing House, 2001. Kindle Edition.

Narrative of Philip Middleton, 4 August 1696, TNA/PRO/CO, CO323/3 f. 114.

Nutting, P. Bradley. "The Madagascar Connection: Parliament and Piracy, 1690–1701." *American Journal of Legal History* 22, no. 202 (1978).

O'Malley, Gregory. *Final Passages: The Intercolonial Slave Trade of British America, 1619–1807*. Chapel Hill: University of North Carolina Press, 2011. Kindle Edition.

Parker, Barry. *The Physics of War: From Arrows to Atoms.* Amherst, NY: Prometheus Books, 2014. Kindle Edition.

Preston, Diana. *A Pirate of Exquisite Mind: The Life of William Dampier: Explorer, Naturalist, and Buccaneer.* New York: Berkley, 2005.

Qaisar, Ahsan J. *The Indian Response to European Technology and Culture (A.D. 1498–1707).* New York: Oxford University Press, 1982.

Rediker, Marcus. *Between the Devil and the Deep Blue Sea: Merchant Seamen, Pirates and the Anglo-American Maritime World, 1700–1750.* Cambridge: Cambridge University Press, 1989.

Robins, Nick. *The Corporation That Changed the World: How the East India Company Shaped the Modern Multinational.* London: Pluto Press, 2012. Kindle Edition.

Rodger, N. A. *The Command of the Ocean: A Naval History of Britain, 1649–1815.* New York: W. W. Norton & Company, 2005.

Steele, Brett D. "Muskets and Pendulums: Benjamin Robins, Leonhard Euler, and the Ballistics Revolution." *Technology and Culture* 35, no. 2 (1994), 348–82.

Stern, Philip J. *The Company-State: Corporate Sovereignty and the Early Modern Foundations of the British Empire in India.* Oxford: Oxford University Press, 2012.

————. "A Politie of Civill & Military Power: Political Thought and the Late Seventeenth-Century Foundations of the East India Company-State." *Journal of British Studies* 47, no. 2 (2008), 253–83.

Subrahmanyam, Sanjay. "Persians, Pilgrims and Portuguese: The Travails of Masulipatnam Shipping in the Western Indian Ocean, 1590–1665." *Modern Asian Studies* 22, no. 3 (1988), 503.

The Trials of Joseph Dawson, William Bishop, Edward Forseth, James Lewis, William May, and John Sparkes for Several Piracies and Robberies by Them Committed. London: John Everingham, 1696.

Thomas, James H. "Merchants and Maritime Marauders." *The Great Circle* 36, no. 1 (2014), 83–107.

Truschke, Audrey. *Aurangzeb: The Life and Legacy of India's Most Controversial King.* Stanford, CA: Stanford University Press, 2017.

Turley, Hans. *Rum, Sodomy, and the Lash: Piracy, Sexuality, and Masculine Identity.* New York: New York University Press, 1999. Kindle Edition.

Van Broeck, Adrian. *The Life and Adventures of Captain John Avery.* Los Angeles: The Augustan Reprint Society, 1980.

Various. *Privateering and Piracy in the Colonial Period: Illustrative Documents.* Kindle Edition.

Woodard, Colin. *The Republic of Pirates: Being the True and Surprising Story of the Caribbean Pirates and the Man Who Brought Them Down.* New York: Houghton Mifflin Harcourt, 2007. Kindle Edition.

Wright, Arnold. *Annesley of Surat and His Times.* London: Melrose, 1918.

Yafa, Stephen. *Cotton: The Biography of a Revolutionary Fiber.* New York: Penguin, 2006.

Zacks, Richard. *The Pirate Hunter: The True Story of Captain Kidd.* New York: Hachette Books, 2003.

Index

Abraham, 123, 151
Abu Ghraib, 42
A Coruña (The Groyne), 68, 75–76,
  79–81, 85, 93, 103, 111, 113, 126,
  133, 227
Adams, Henry, 59, 195, 197, 205, 243–44
  wife of, 204, 205, 243–44, 250–51
Adams, John Quincy, 24
Adventure Galley, 253
Afghanistan, 67
Africa, 111, 118, 125, 187, 188
Agra, 47–48, 53, 54, 64, 65, 67, 124
Ajanta Caves, 32
Akbar the Great, 37, 136, 137
Al-Biruni, 35–36
Alexander the Great, xi, 33–34
Algiers, 39–42
al-Qaeda, 248
American colonies, 204–5
American Revolution, 91
American Weekly Mercury, 27, 160
Amity, 113–14, 128, 144, 147–48, 161
Ammurapi, King, 23
Anatomy of Exchange-Alley (Defoe), 105–6
Anglo-Spanish War, 46
Annesley, Samuel, 107–8, 110, 150, 164–68,
  172, 176, 198, 236, 254
  death of, 254
  firing of, 254
  plan to make East India Company the
    protector of Indian ships, 198–201, 206,
    247, 253, 255
Antarctic Circumpolar Current, 118–19
Apple, 105

Arabian Sea, 124, 145
articles of agreement, 88–91, 129, 160,
  184, 195
Ascension, 188
Asir Mountains, 121
Augustine, St., xi
Aurangzeb, 37–38, 64–67, 107, 109,
  110, 113, 129, 151, 162, 184, 211,
  253–54
  Annesley and, 165–66, 168
  and arrests of Every's crew, 208
  and criminal trials of Every's crew, 223,
    225, 227, 236
  East India Company and, 107, 109, 110,
    131, 169, 189, 192
  East India Company as protector of
    ships of, 198–201, 206, 211–12, 247,
    253, 255
  Every and, 166
  Fath Mahmamadi and, 165–66
  granddaughter of, 136, 152–54, 157–58,
    162–63, 245–46, 250–52
  Gunsway attack and, 168, 169, 170, 172–74,
    176, 198
  Khan and, 155–56, 157, 158, 161–62, 165,
    170, 189–90, 198
  sisters of, 137
Australia, 60
Avery, John and Anne, 17

Bab-el-Mandeb, 124, 128, 145
Babur, 64, 67
Baer, Joel, 100, 132
Bahamas, 187, 193–97, 203–5, 231

INDEX

ballads, 97–98, 100, 152, 157–58, 162, 173, 232, 214
 Every Verses, 95, 96–101, 102–3, 265n
 transition between song and print, 97
Bangladesh, 67
Bank of England, 56
Barbados, 112
Barbary pirates, 40–42, 44, 55, 57
Battle of Plassey, 200, 254
Baudhayana sutra, 34–35
Beagle, 60
Beg, Usher, 165
Bellamy, Samuel "Black Sam," 7, 16
Bermuda, 113, 114
Bernier, François, 137, 139
Bezos, Jeff, 51
bin Laden, Osama, 179
Bishop, William, 59, 72, 80
 trial of, 220, 223, 227, 233, 239, 240
Black and British (Olusoga), 116
Blackbeard, 7, 8, 16, 116
Blackborne, Robert, 176–77, 206–7
Black Caesar, 116
Board of Trade, 57, 175, 177, 179, 203
Bohun, George, 175
Bolan Pass, 30–31, 66
Bombay, 145, 155, 174, 200
 assaults on, 109, 166–67, 169, 174
 Bombay Castle, 109, 167, 169, 170, 174, 189–90, 221, 255
 coins minted in, 168, 192
 East India headquarters at, 54, 107–9, 130, 151, 168–69, 174, 202
Brahui mountains, 30, 31, 66
Braudel, Fernand, 36–37
Brazil, 118
Britain, 249
 India and, 8, 9, 38, 154–55, 200, 202, 211, 221
 navy of, see Royal Navy
 piracy denounced by, 208, 211–16, 221–22, 225–26, 231, 236, 241–42, 247, 254, 255
 Spain and, 43
 West Country, 13, 16
 wool trade in, 172–73, 202
broadsides, 97, 98, 100, 101
Bromley, J. S., 94
Bronze Age, 20, 23
Buckland Abbey, 44

Buffett, Warren, 51
Burgess, Douglas R., Jr., 43, 44, 212, 226

Calico Jack, 7
cannons, 2–3, 5
 explosion of, 3, 6
 explosion on Gunsway, 3–4, 6, 151, 152, 158, 249
Cape of Good Hope, 112, 113, 117, 118
Cape St. John, 145, 146
Cape Verde, 111–12, 114, 126
capitalism, 14, 248
Caribbean, 56, 73, 113, 253, 254
Carlos II, King, 56
Catherine of Braganza, 107
Cayenne, 187
Central America, 44
Chach, 31
Charles II, 56–60, 67, 103, 114, 239
 Every's crew tried for mutiny on, 226–37, 238
 Every's mutiny on, 76, 79–87, 92–93, 96, 97, 99, 100, 112, 133, 171, 176, 195
 rechristened the Fancy, 87
Charles II, King, 107
Child, Josiah, 105–7, 110
China, 32, 122
Chinese New Year, 122
Christians, 64, 127
City of God, The (Augustine), xi
class mobility, 75, 245–46
Clive, Robert, 200
coffee, 121
coins:
 found in Dann's coat, 207, 243
 minted in Bombay, 168, 192
Committee on Trade and Plantations, 103
Comoro Islands, 126, 130, 230
concubines, 135, 138, 158–59
Constitution, U.S., 92
Cooke, Thomas, 110
corporations, 51, 89, 93, 171–72, 174, 249
 stock in, 52–53, 105
cotton, 32–34, 50, 53, 104, 121, 124
 dyes and, 32–34
 English backlash against imports of, 172–73, 202
Creagh, David, 59, 80, 82–83, 86, 228–29, 233

INDEX

da Gama, Vasco, 45, 118
Dampier, William, 59–60, 72, 73, 74,
    82, 103–4
Danish privateers, 118
Dann, John, 59, 127, 143, 144, 146, 183,
    187, 205–7
    Adams's wife encountered by, 243–45
    arrest of, 207, 243
    coins found in coat of, 207, 243
    confession of, 215
    trial of, 218, 222, 229–31
Dara, Prince, 64–66, 137
Darwin, Charles, 60
Dawson, Joseph, 93, 112, 159, 184, 185
    trial of, 220, 221, 223, 227, 238, 239
de Clisson, Jeanne-Louise, 25–26, 29
de Clisson, Olivier, 25
Defoe, Daniel, 17, 59, 105–6, 107, 153,
    154, 173
Delhi, 36, 63, 64, 66, 124, 166
Devon, 13, 16, 44, 252
Devonshire, 17, 44, 74, 99, 130, 245
    piracy and, 15–16
Din-I Ilahi, 37
disease and illness, 71, 75
    at Bombay, 109
    dysentery, 69–72
    scurvy, 71
    typhus, 217–18
    venereal, 71
doctors, 71
Dolphin, 127–28, 144
Dove, 56, 58, 59, 82
Drake, Francis, 16, 17, 44, 52, 75, 116, 148,
    154, 171, 185, 211
Druit, Thomas, 59, 79–81, 228,
    232, 233
Dunfanaghy, 205
Dutch, 40, 155, 202, 211
dyed fabrics, 32–34
dysentery, 69–72

East India Company, 1, 46, 47, 50–54, 56,
    57, 88–89, 104–10, 126, 130, 136, 152,
    161, 162, 185, 200–202, 207, 211, 253,
    254, 255
    and arrests and trials of men from Every's
        crew, 207–8, 236
    Aurangzeb and, 107, 109, 110, 131, 169,
        189, 192

and backlash against cotton imports,
    172–75, 202
Battle of Plassey and, 200, 254
Board of Trade and, 175, 177, 179
Bombay headquarters of, 54, 107–9, 130,
    151, 168–69, 174, 202
charter of, 51, 53, 106, 107, 109
Child in, 105–7
coins printed by, 168, 192
corruption in, 105–7, 109–10, 174
Every and, 130–32, 149
Fath Mahmamadi and, 164–65, 168, 169
Gunsway affair and, 164–69, 170–72,
    174–79, 198–99
increased naval power of, 206, 247
and Khan's meeting with Gayer, 189–92
and pirates of the Red Sea, 108–9
power of, 106
protection to Indian ships provided
    by, 198–201, 206, 211–12, 247, 253, 255
stock in, 51–53, 104–5, 110, 173
Surat factory of, 1, 46, 54, 107–10, 151,
    164–69, 176, 198–201, 253–55
Edward I, King, 43
Egypt, 19–22
Elizabeth I, Queen, 44, 51, 53
enemies of all mankind, 41–42, 44, 179, 241,
    246, 248
England, *see* Britain
English Channel, 16, 25–26, 29, 41, 74
Enlightenment, 26, 213
Europe:
    India and, 49
    Orientalist literature in, 48
Everingham, John, 225, 242
Every, Henry, 13–18, 245–53
    Adams's wife and, 243–45, 250–51
    armada commanded by, 129, 132–33,
        143–45, 147
    Aurangzeb and, 166
    Aurangzeb's granddaughter and, 152–54,
        157–58, 162–63, 245–46, 250–52
    birth of, 16–18, 37, 40
    charisma of, 133
    childhood of, 17
    criminal trials for crew of, *see* trials of
        Every's crew
    death of, 18, 252, 270n
    disappearance of, 245
    East India Company and, 130–32, 149

Every, Henry *(cont.)*
  ethics in actions of, 112, 133
  Fath Mahmamadi and, 145–48
  first years of naval career, 44
  getaway of, 183–88
  Gunsway gold and, 185
  Gunsway sighted by, 146–49, 157
  Gunsway treasure and, 151–52
  Khan and, 63
  last trace of existence of, 245
  letter written at Johanna Island by,
      131–34, 148–49
  manhunt and reward for, 163, 177–79,
      183–88, 194, 195, 205–7, 219, 255
  memoir of, 17–18
  and models of piracy, 44
  mythology of, 101, 128, 212, 224, 245–46,
      250–51
  name and aliases of, 16–18, 186–87, 206
  in Nassau, 193–97, 205, 231
  notoriety of, 75, 94–95
  Royal Navy joined by, 13, 15, 18, 39–41, 55
  ship and crew of, *see Fancy*
  slaves purchased by, 186–88
  as slave trader, 55, 56, 113, 116
  on Spanish Expedition, 60–61, 66–67
  Spanish Expedition mutiny led by, 76,
      79–87, 92–93, 96, 97, 99, 100, 112, 133,
      171, 176, 195
  Tew and, 114, 128, 129, 132, 147
  Tew's death and, 148
  Trott and, 194–97
  Van Broeck's biography of, 39–40,
      152–54, 245, 250–51, 267n
  verses on, 95, 96–101, 102–3, 265n
  wealth of, 75, 185
Execution Dock, 238–42, 243, 250, 253
Exquemelin, Alexandre, 93, 160
Exuma, 193

fame, 74–75
Fancy, 9, 34, 88–95, 99, 101, 111–19, 128–29
  alterations made to, 116–18
  arrests of crewmembers of, 207–8, 212
  articles of agreement aboard, 88–89
  in Bahamas, 194–97
  at Cape Verde, 111–12, 114, 126
  Charles II rechristened, 87
  criminal trials for crew of, *see* trials of
      Every's crew

Danish privateers and, 118
  dispersal of crew of, 203–8
  Every's abandonment of, 195–97, 203
  Fath Mahmamadi and, 145–48, 164–65,
      168, 169, 215
  at Fernando Po, 116–18
  getaway of, 183–88
  at Guinea, 114–15, 133
  guns of, 111
  Gunsway battle with, 1–7, 9, 151–56,
      157–63
  Gunsway encounter narratives, 152–56,
      157–63
  Gunsway officers tortured by, 159–60,
      163, 168
  Gunsway treasure divided among crew of,
      183–85
  Gunsway women assaulted by, 160–63,
      166, 214, 215, 241
  hunt for crew of, 177–79
  Low and, 89
  at Madagascar, 83, 112–13, 118, 119, 120,
      126, 147
  mosque at Maydh demolished by crew of,
      126–27, 130, 132, 133, 160, 215, 248
  Red Sea plans of, 113–14, 126, 127, 131,
      144, 163
  shipworms and, 117–18
  slaves on board, 115–16, 133, 186–88,
      195, 248
  speed of, 1–2, 57, 81, 117, 118, 129, 131,
      144, 149, 188
  West Wind Drift and, 118–19
  wreck of, 203–4
Fath Mahmamadi, 145–48, 164–65, 168, 169,
      183–84, 215
faujdars, 199
East India Company as protector of Indian
      ships, 198–201, 211–12, 247, 253, 255
Fernando Po, 116–18
feudalism, 14, 51
Fleet, John, 175
Forsyth, Edward, 59, 80
  trial of, 220, 223, 227, 232–33, 236–37, 238,
      240–42
Fort Dauphin, 126
Foster, William, 48
France, 126, 170, 186, 196–97
  Revolution in, 24, 91
Franz Ferdinand, Archduke, 249

Gacy, John Wayne, 26
Ganj-i-Sawai, see Gunsway
Garsia Cassada, Andres, 58
Gates, Bill, 51
Gayer, John, 130–31, 161–63, 167–68, 172, 174, 199, 201, 208, 211–12, 221, 222, 225, 236, 255
  Khan's meeting with, 189–92
General History of the Pyrates, A (Johnson), 43, 91, 92, 152, 246, 252, 265n, 270n
Ghaffar, Abdul, 145, 146, 164, 167, 199
Gibson, Charles, 58, 80–81, 83–87, 99, 112, 184, 227–29, 233, 234
Glorious Revolution, 106
gold, 34, 50, 124, 159, 185, 252
Gravet, Joseph, 59, 81, 84–87, 228, 234
Great Britain, *see* Britain
Greece, 20
Groyne, The, *see* A Coruña
Guantanamo Bay, 42
Guinea, 114–16, 133, 205
Gujarat, 65
Gulf of Aden, 124, 126–28, 144–45, 147–48
Gulliver's Travels (Swift), 59
gunpowder, 3
Gunsway (Ganj-i-Sawai), 120–21, 124, 135–36, 146–49, 248
  Aurangzeb's response to attack on, 168, 169, 170, 172–74, 176, 198
  in battle with the *Fancy*, 1–7, 9, 151–56, 157–63
  Bombay siege and, 166–67
  cannon explosion on, 3–4, 6, 151, 152, 158, 249
  collapse of mainmast on, 5–7, 152, 158, 249
  East India Company and, 164–69, 170–72, 174–79, 198–99
  Every's sighting of, 146–49, 157
  Khan's account of events on, 155–56, 157–62
  narratives of raid on, 152–56, 157–63
  officers tortured by Every's crew, 159–60, 163, 168
  pact following attack on, for East India Company to protect Indian ships, 198–201, 206, 211–12, 247, 253, 255
  survivors of, at Surat, 166
  treasure on, 151–52, 159–60, 183–85

trials of crewmembers involved in attack on, *see* trials of Every's crew
women on board, 135–40, 152–54, 158–59
women on board assaulted by Every's crew, 160–63, 166, 214, 215, 241

Habermas, Jürgen, 213
Hagar, 123
Haiti, 58
hajj (pilgrimage to Mecca), 113, 121–24, 126, 135–36, 138–39, 150, 151, 166, 202, 255
Hamond, Walter, 125–26
harems, 138–39, 251
Hawkins, William, 46–51, 53–54, 63, 107
Hector, 45–46
Hedges, Charles, 218–21, 225–27, 235, 236, 238, 239
Herodotus, 32
High Court of Admiralty, 103–4, 213, 214, 216
Hinduism, 36–37, 64, 67
History of Civilization, A (Braudel), 36–37
Hog Island, 194, 197
Holt, John, 218, 219, 225, 226, 230–37
Homer, 21
hostis humani generis (enemies of all mankind), 41–42, 44, 179, 241, 246, 248
Houblon, Isaac, 175–76, 219
Houblon, James, 56–58, 68, 75–76, 84, 88, 96, 102–3, 114, 133, 175–76, 184, 203, 205, 206, 219, 249
  and trial of Every's crew, 226, 227
Houblon, John, 56, 57, 219
House of Commons, 109
Humphries, Captain, 80–81

iconoclast, use of word, 36
impressment, 13–15
India, Indian subcontinent, 31–32
  Battle of Plassey and, 200, 254
  Britain and, 8, 9, 38, 154–55, 200, 202, 211, 221
  cotton of, *see* cotton
  Europe and, 49
  gold in, 50
  Mughal dynasty in, 8, 9, 36–38, 50, 62–63, 108, 113, 157, 249, 254
  trade and, 32–35, 50, 54; *see also* East India Company
  women in, *see* women

INDEX

Indian Ocean, 119, 150–51, 185, 200
Indonesia, 155
Indus Valley, 31–33
interlopers, 186
Invisible Hook, The (Leeson), 28
Ireland, 205–7, 215, 244, 250, 270n
ISIS, 248
Islam, Muslims, 31–32, 35–36, 64, 151
  Fancy and, 127
  harem culture and, 138
  Hindu culture and, 36–37
  Islamic calendar, 122
  mosque at Maydh burned by Every's
    crew, 126–27, 130, 132, 133, 160,
    215, 248
  pilgrimage to Mecca (hajj), 113, 121–24,
    126, 135–36, 138–39, 150, 151, 166,
    202, 255
  Quran, 123, 138
  traders, 34, 35
Ismail (Ishmael), 123

Jack the Ripper, 26–27
Jacobins, 24
Jahanara, 64, 137
Jahangir, 46–51, 53–54, 63, 136–37, 202
Jamaica, 73
James, 56, 59, 79–81, 228, 232
James II, King, 106
James VI and I, King, 46, 47, 50, 51, 54
Jeane of Bristol, 27
Jefferson, Thomas, 24
jewels, 34, 124, 151, 160, 184
Jiwan, Malik, 66
Johanna Island, Every's letter at, 131–34,
  148–49
Johnson, Charles, 43, 91, 92, 152, 246, 252,
  265n, 270n
Justice Department, U.S., 42

Kaaba, 122, 124
Kangra, 36
Kealy, John, 263n
Keay, John, 50
Khan, Asad, 201
Khan, Ibrahim, 158–59
Khan, I'timad, 165–68, 201
Khan, Khafi, 63–66, 155–56, 157–62, 165, 166,
  168, 169, 170, 189, 198, 201
  Gayer's meeting with, 189–92

Khan, Muqarrab, 47
Khan, Yakut, 191
Kidd, William, 253, 255
King of Pirates, The (Defoe), 153
Knight, John, 58
Konstam, Angus, 43, 264n

Lawson, John, 39–40
Leeson, Peter, 28
letter of marque, 43, 171
Lewis, James, 80
  trial of, 220, 223, 233, 239, 240
Lewis, Theophilus, 96
Libertalia, 246, 252
Life and Adventures of Captain John Avery, The
  (Van Broeck), 39–40, 152–54, 245,
  250–51, 267n
life expectancy, 72
Lind, James, 71
Lisbon, 57
literacy, 73
Littleton, Thomas, 228
l'Ollonais, François, 27
Low, Edward, 27, 89, 160, 161
Lowther, George, 89

Madagascar, 83, 112–13, 118, 119, 120,
  125–26, 147, 185–87, 204, 246, 252
Madras, 54
Madrid, 68
Mahmud of Ghazna, 36, 263n
Malagasy people, 125–26
Manson, Charles, 26
Manucci, Niccolao, 137, 139
Marx, Karl, 248
Mary II, Queen, 170
May, William, 59, 80, 83–87
  trial of, 220, 223, 227–31, 233–36, 238–40
Maydh, burning of mosque at, 126–27, 130,
  132, 133, 160, 215, 248
Mayflower, 125, 135
Mecca, 31, 122–23, 151
  pilgrimage to (hajj), 113, 121–24, 126,
    135–36, 138–39, 150, 151, 166, 202, 255
Medinet Habu, 19, 22
media, 26–29, 74, 94–95
  broadsides, 97, 98, 100, 101
  newspapers, 97
  transition between song and print, 97
  see also ballads

medical research, 71
Mediterranean, 20–21, 23, 40, 41
Memoirs of the Mughal Court (Manucci), 137, 139
Middleton, Philip, 59, 114, 115, 128, 129, 144, 163, 185, 187, 195–97, 203–5, 215, 253
  trial of, 218, 222, 229, 231, 232, 253
mlecchas, 31
Mocha, 121, 143–44
money laundering, 186–87, 252
Monroe, James, 24
mortality rates for voyages, 72
Mues, William, 128
Mughal Empire, 8, 9, 36–38, 50, 62–63, 108, 113, 157, 249, 254
  see also India, Indian subcontinent
Muhammad, 31, 35, 122–23, 151
Muhammad bin Qasim, 31
Muhammad Ghuri, 36
Mukherjee, Soma, 138
Mumtaz Mahal, 62
Murad, Prince, 64
Muslims, *see* Islam, Muslims
Mycenaean age, 20

Nassau, 187, 193–97, 203–5, 231
Newgate Prison, 217–18, 225–26, 229, 239
New Providence, 187–88, 193–97
Newton, Henry, xi, 216, 219, 221–23, 225, 231–32, 235, 246, 254–55
  opening statement of, 221–22, 269n
Newton Ferrers, 16–17
New Voyage Round the World, A (Dampier), 59–60, 74
Nigeria, 116
Nile Delta, 19–22
9/11 attacks, 42
Nine Years' War, 170

O'Byrne, Don Arturo, 56
Odysseus, 26
Odyssey, The (Homer), 21
oil, 124
Old Bailey, 217–23, 225–37, 253, 255
Olusoga, David, 55–56, 116
Orientalist literature, 48

Pakistan, 30–32, 67
pamphleteers, 26–29, 74, 95, 152, 162, 212, 214, 240

Paradise Island, 194
Parliament, 15, 106–7, 109–10
Pascal, Blaise, 6
Pearl, 128, 144, 183, 188
pepper, 121
Perim, 128, 129, 143, 145
Philip VI, King, 25
Phillips, John, 89
Phillips, Thomas, 55, 56
pilgrimages, 123–24
  to Mecca (hajj), 113, 121–24, 126, 135–36, 138–39, 150, 151, 166, 202, 255
pirates, 7–8, 16, 151, 172, 247–48
  ages of, 264n
  American colonies and, 204
  articles of agreement of, 88–91, 129, 160, 184, 195
  Barbary, 40–42, 44, 55, 57
  British government's denunciation of, 208, 211–16, 221–22, 225–26, 231, 236, 241–42, 247, 254, 255
  codes of behavior of, 88–93
  compensation plans of, 92, 93
  democratic principles of, 91, 93
  Devonshire and, 15–16
  distribution of profits among, 89, 184
  dominant historical narrative of, 212–14, 223, 242
  East India Company and, 108–9
  East India Company as protector of Indian ships from, 198–201, 211–12, 247, 253, 255
  egalitarian and populist ethos of, 94, 151, 246–48, 265n
  as enemies of all mankind, 41–42, 44, 179, 241, 246, 248
  execution of, 241
  flags of, 21
  global population of, 128
  golden age of, 26, 29, 254
  Hollywood representations of, 70
  insurance policies of, 93
  law and trials for crimes of, 214
  at Madagascar, 126
  as political pioneers, 93–94
  privateers vs., 17, 42–44, 60
  quartermasters, 92–93
  in Red Sea, 2, 108–9, 113–14, 128, 131, 150, 253, 255
  reputations of, 26, 28–29

pirates *(cont.)*
  Sea Peoples, 19–25
  self-identity in, 21
  separation of powers created by, 92, 93
  slaves and, 115–16
  Somali, 124
  terrorism and, 24, 25
  torture and atrocities committed by,
    27–29, 154–55, 159–60, 248, 261n
  utopia of, 246
Plassey, Battle of, 200, 254
Port Royal, 73
Portsmouth Adventure, 127–28, 144
Portugal, 45–47, 54, 57, 107, 111–12, 115, 126,
  136, 202
press, *see* media
privateers, privateering, 43–44
  as career path, 72–73
  financial rewards from, 75
  Hollywood representations of, 70
  pirates vs., 17, 42–44, 60
  see also sailors
Privy Council, 57, 102–3
prostitutes, 71, 73
public sphere, 213
Puritans, 125

quartermasters, 92–93
Quedagh Merchant, 253
Queen Anne's Revenge, 116
Quran, 123, 138

Rahimi, 136–37
Rahiri, 155, 189
Rajputs, 64
Raleigh, Walter, 16, 17
Ramses III, 19–23
Razzak, Abdur, 189, 190
Rediker, Marcus, 94, 264n
Red Sea, 45, 121, 124, 127, 143
  Fancy's plans for, 113–14, 126, 127, 131,
    144, 163
  pirates in, 2, 108–9, 113–14, 128, 131, 150,
    253, 255
  Tew and, 114, 128, 129
religion, 64, 123
  see also Islam, Muslims
Resolution, HMS, 39–40
Réunion, 185–86, 205

Rhode Island, 204
Roberts, Bartholomew, 89–91, 93
Robespierre, Maximilien, 24, 25
Robin Hood ballads, 100
Robins, Nick, 109–10
Robinson Crusoe (Defoe), 59
Roe, Thomas, 54
Roebuck, HMS, 60
Rogers, Woodes, 252
Roshanara Begum, 137
Royal African Company (RAC),
  55–56, 115
Royal Navy, 17, 40, 42, 58, 71, 89, 133,
  170–71
  Every's joining of, 13, 15, 18, 39–41, 55
  press-gangs and, 13–15
  wages paid by, 185
Royal Society, 60

sailors:
  atrocities committed by, 154–55
  life aboard ship, 68–74
  literacy rates of, 73
  mortality rates of, 72
  see also privateers, privateering
Sailors Advocate, The, 14
Saint-Augustin Bay, 119, 125, 126
Saudi Arabia, 122, 138
scientific research, 71
Scotland, 206
scurvy, 71
Sea Flower, 205–6
Sea Peoples, 19–25
seaweed, 118
sensationalism, 26–29
September 11 attacks, 42
serial killers, 26–27
Seventh Son, 56
sexual experiences, 73
sexually transmitted diseases, 71
Shah Jahan, 62–64, 66, 67
ships:
  life aboard, 68–74
  seaweed and, 118
  shipworms and, 117–18
  see also sailors
shipworms, 117–18
Shuja, Prince, 64
Sindh, 31

slavery, slave trade, 41, 55, 76, 115, 126
  in American colonies, 205
  Drake and, 116
  Every as slave trader, 55, 56, 113, 116
  female slaves, 138
  pirates and, 115–16
  Royal African Company and, 55–56, 115
  slaves on board the *Fancy*, 115–16, 133,
    186–88, 195, 248
  Spanish Expedition crew and, 76, 80
Somalia, 126
Somali pirates, 124
songs, *see* ballads
Son of Sam, 26
Spain, 43, 44, 115, 193
Spanish Armada, 16
Spanish Expedition Shipping venture,
    56–61, 68–76, 88, 89, 114, 146, 167,
    175–76, 186, 219
  Charles II in, 56–60, 67, 103
  departure from England, 58, 264n
  Every in, 60–61, 66–67
  Every's crew tried for mutiny on,
    226–37, 238
  Every's mutiny on the *Charles II*, 76,
    79–87, 92–93, 96, 97, 99, 100, 112, 133,
    171, 176, 195
  James in, 56, 59, 79–81, 228, 232
  in legal imbroglio, 102–4, 109
  return of remainder of crew, 103
  slavery fears and, 76, 80
  wages paid in, 57–58, 103, 185
  wives of crewmembers of, 102
Spanish galleons, 56, 113
Sparkes, John, 59, 162, 185
  trial of, 220, 223, 231, 233, 236,
    239–41
Spice Islands, 46, 52, 155, 211
spices, 34, 121, 124, 159
Stern, Philip, 171, 200
Strabo, 33, 34, 262n
strikes, labor, 73
Strong, John, 58
Successful Pyrate, The, 246
Suez Canal, 124, 143
Sufis, 64
Surat, 1, 2, 45–48, 108, 120, 122, 143, 145,
    149, 150–51, 155–56, 157, 174, 183, 189,
    211, 250

East India Company factory at, 1, 46, 54,
    107–10, 151, 164–69, 176, 198–201, 253–55
  Gunsway survivors at, 166
Susanna, 144, 183, 188
Swift, Jonathan, 59
syphilis, 71

Taj Mahal, 47, 62, 67
Tapti River, 45, 150
terrorism, 23–25, 42
  meanings and use of word, 24–25
  modern, 24–25
  in Mughal reign over India, 36–37
  9/11 attacks, 42
  piracy and, 24, 25
Tew, Thomas, 113–14, 128, 129, 132, 144,
    146, 147, 204
  death of, 147–48, 161
Thames, 58, 60, 67, 185, 204, 240
Thar Desert, 31–32
Theatre Royal, 246
Thoreau, Henry David, 117
"Through All the Fates" (Thoreau), 117
trade, 32, 34
  in cotton, *see* cotton
  East India Company and, *see* East India
    Company
  India and, 32–35, 50, 54
  Muslim traders, 34, 35
travel, 74
travel writing, 59–60, 74
Treaty of London, 46
trials of Every's crew, 246
  acquittals in, 223, 224–26, 239
  for mutiny, 226–37, 238
  Newton's opening statement in,
    221–22, 269n
  for piracy, 212–16, 217–23, 224–26, 254–55
  sentencing and executions in, 237,
    238–42, 243, 249, 250
Tripoli, 41
Trott, Nicholas, 193–97, 203–5, 215, 231
turtles, 188
Tutankhamun, 19
Tyler, John, 6
typhus, 217–18

Vagabond Act, 15
vagabonds, 14–15

INDEX

vagrancy, 15
Van Broeck, Adrian, 39–40, 152–54, 245, 250–51, 267n
venereal diseases, 71

Wake, Thomas, 128
warfare, asymmetric, 23–25
Wesley, John, 107
West Country, 13, 16
West Indies, 68, 76, 112–13, 116, 187, 196
West Wind Drift, 118–19
William III, King (William of Orange), 106, 170, 172, 177, 193, 255

women, 135–40
  on *Gunsway*, 135–40, 152–54, 158–59, 166
  in harems, 138–39, 251
  sexual violence against, 160–63, 166, 214, 215, 241
Woodard, Colin, 197
wool trade, 172–73, 202
World War II, 5
Wright, Arnold, 108, 164, 166, 169, 201

Yemen, 121
Yoo, John, 42

Zeb-un-Nissa, Princess, 138

WONDERLAND
How Play Made the Modern World

Play has always been more important to innovation and creativity than most people realize. Just as *How We Got to Now* investigated the secret history behind everyday objects, this vivid examination of the power of play and delight offers a surprising history of popular entertainment. Roving from medieval kitchens and ancient taverns to casinos, theaters, computer labs, and shopping malls, Steven Johnson locates the cutting edge of innovation wherever people are working hardest to keep themselves and others amused.

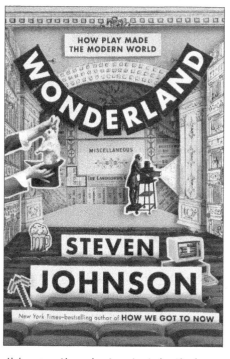

These wonderlands of amusement did more than just entertain their patrons, Johnson argues. They also directly contributed to economic and social revolutions that transformed the modern world.

"A house of wonders itself . . . *Wonderland* inspires grins and well-what-d'ya-knows." **—The New York Times Book Review**

"A rare gem . . . Our illogical, enduring fascination with play remains one of life's great mysteries. That is precisely what makes the subject so fascinating, and *Wonderland* such a compelling read." **—The Washington Post**

"Johnson's prose is nimble, his knowledge impressive. . . . *Wonderland* is original and fun, as well it should be, given the subject." **—San Francisco Chronicle**

FARSIGHTED

How We Make the Decisions That Matter the Most

Plenty of books offer useful advice on how to get better at making quick-thinking, intuitive choices. But what about more consequential decisions, the ones that affect our lives for years or centuries to come? Steven Johnson examines how we can better make the decisions that alter the course of a life, an organization, or a civilization. Deploying the beautifully crafted storytelling and novel insights that Johnson's fans know to expect, he draws lessons

from cognitive science, social psychology, military strategy, environmental planning, and great works of literature. Ranging from how Charles Darwin weighed his decision to get married to our modern leaders' approaches to climate change and artificial intelligence forecasting, *Farsighted* explains how we can address big choices of all kinds more intentionally and effectively.

The Ghost Map: The Story of London's Most Terrifying Epidemic—and How It Changed Science, Cities, and the Modern World

A *New York Times* Notable Book

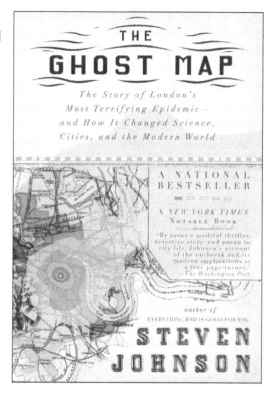

A riveting page-turner about a real-life historical hero, Dr. John Snow. In the summer of 1854, London has just emerged as one of the first modern cities in the world. But lacking the infrastructure—garbage removal, clean water, sewers—necessary to support its rapidly expanding population, the city has become the perfect breeding ground for a terrifying disease no one knows how to cure. As the cholera outbreak takes hold, a physician and a local curate are spurred to action—and ultimately solve the most pressing medical riddle of their time.

Johnson illuminates the intertwined histories and interconnectedness of the spread of disease, contagion theory, the rise of cities, and the nature of scientific inquiry.

"Thrilling."   **—GQ**   "Vivid."   **—The New Yorker**

"Marvelous." **—The Wall Street Journal**   "Fascinating."

—The New York Times Book Review

T288-0413

How We Got to Now
Six Innovations That Made the Modern World

In this illustrated volume, Steven Johnson explores the history of innovation over centuries, tracing facets of modern life (refrigeration, clocks, and eyeglass lenses, to name a few) from their creation by scientists, engineers, hobbyists, amateurs, and entrepreneurs to their unintended historical consequences. Filled with surprising stories of accidental genius and brilliant mistakes, *How We Got to Now* investigates the secret history behind contemporary life.

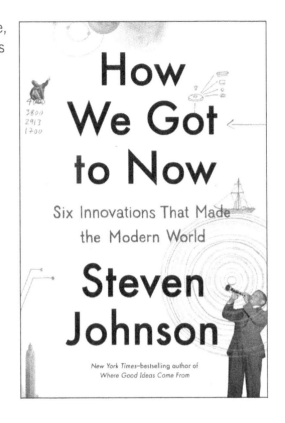

Accompanied by a major six-part television series on PBS, *How We Got to Now* is the story of collaborative networks building the modern world, written in the provocative, informative, and engaging style that has earned Johnson fans around the globe.